Prisoner

Prisoner

My 544 Days in an Iranian Prison—
Solitary Confinement, a Sham Trial,
High-Stakes Diplomacy, and the
Extraordinary Efforts It Took
to Get Me Out

Jason Rezaian

HARPER LUXE

An Imprint of HarperCollinsPublishers

HarperCollins books may be purchased for educational, business, or sales promotional use. For information please e-mail the Special Markets Department at SPsales@harpercollins.com.

FIRST HARPERLUXE EDITION

ISBN: 978-0-06-288826-6

HarperLuxe™ is a trademark of HarperCollins Publishers.

Library of Congress Cataloging-in-Publication Data is available upon request.

19 20 21 22 23 ID/LSC 10 9 8 7 6 5 4 3 2 1

For my wife, my brother, and my mom . . . My heroes

1
Arrest

It was a Tuesday afternoon in Tehran in the middle of Ramadan. My wife, Yeganeh, also a journalist, and I literally had our bags packed to leave for the United States. Her U.S. immigration papers had come through, and we had tickets for a flight that Friday night, with plans to travel to America and be away from work for a couple of months.

I had noticed, walking around town, that contrary to past years, some restaurants were open during daytime hours this Ramadan. I took it as a small sign of pragmatism on the part of the government. In the past fasting had been enforced and restaurants serving during daylight hours were fined or shuttered. Now they weren't. Progress? Maybe.

So for the last piece I would work on for the *Washington Post* before leaving Iran for our extended break, I went to interview a guy who had an American-style diner in Tehran. It was the kind of story I loved doing. Food, and eating it, have always been a passion of mine and I had recently found ways to bring gastronomy into my coverage of Iran, but it was the incongruity of the various elements that attracted me. This sort of piece was my specialty.

In the middle of our conversation about serving chili dogs and sliders during Islam's holiest month, around three o'clock in the afternoon, I got a frantic call from my wife.

"Somebody's trying to destroy our lives," she told me. "You need to come home now."

I rushed home from the east to the west of the vast capital and when I got there, she was crying hysterically. She showed me an email she'd received, written in Farsi, which I couldn't read.

Pay us 10,000,000 toman by tomorrow at 3:00 p.m., she translated, *or we'll expose you for the whore that you are.*

It seemed so strange. It was obviously directed at her—they were making reference to her personally. It was from an email address we didn't recognize. The amount they claimed they wanted was only about three thousand dollars.

"There are people trying to destroy us," she said.

It wasn't the only strange Internet activity we'd seen recently. A few days before, I'd gotten a phishing email that appeared to be from a photographer and dual Iranian and U.S. national I worked with frequently who did a lot of freelance work for the *Washington Post* and had become something of a semiofficial photographer for President Rouhani, traveling with him and his entourage extensively.

And just the day before, an Iranian source of mine who was very close to Iran's foreign minister, Mohammad Javad Zarif, had sent me a text saying that he had received an email from a Gmail account that "looked like yours but wasn't yours, with a link. Be careful."

These phishing emails had started appearing just before I went to Vienna, where I'd covered the latest round of the nuclear negotiations. I'd written a piece for the *Post* saying that the talks hadn't come to a final agreement but were being extended for another six months—everyone knew that neither side wanted the diplomatic process to break down.

I called our friend Reza, who helped us with IT matters. I thought he'd be able to figure out what was going on by checking programs on our computers that I didn't know about. He came to the apartment, and he looked at the activity of our Gmail accounts. Over the previous few hours, we discovered, both of our accounts had been accessed from a server in Russia, multiple times.

"Look, this could just be people trying to get financial information from you guys," Reza said. He changed the passwords. I thought everything was fine at that point.

"We should just go to the airport and leave tonight," Yegi said.

"This is not something that we should be overly concerned about," I said. "Nothing's going to happen. You can go to Dubai and wait for me if you really want to, but I need to finish this Ramadan story."

Around eight or eight thirty, we called two cabs, one for ourselves and one for Reza. Because gasoline is cheap in Iran, taxis are also inexpensive and readily available twenty-four hours a day. In every neighborhood, there are several taxi shops, so within a couple minutes, you can usually get a car.

Yegi and I were headed to a surprise party for her mom's fifty-sixth birthday, a dinner at a cousin's house. We were dressed casually, me in jeans and a pink shirt, Yegi in a beautiful, short blue dress I had just bought for her, and wearing makeup and jewelry.

The doorman called from downstairs when the taxis arrived. We took the elevator down with Reza four floors into the parking garage in the building's basement, where cars arrive.

When the elevator door opened, there was a guy standing there with a gun pointed at me.

There was more than one guy, but I just remember the one with the gun. I was focused solely on him. He wore a gray suit and was very nondescript—shorter than me, with a gut, a comb-over, and a mustache. Straight out of central casting. His gun was a little revolver.

"Rezaian?" he asked.

"Yeah," I said, startled.

He propped the elevator door open with his foot. I went to grab my cell phone—I don't know who I thought I was going to call—but he knocked it out of my hand. It was a very frantic scene. He had a piece of paper that was a warrant for our arrests, although it was unclear which of the many competing seats of authority had ordered it. The men didn't identify themselves. My Farsi was good, but not good enough in that moment to understand completely when he said that we were being arrested.

"Don't make any noise," the man with the gun ordered us. "We're going back to the apartment."

I had no idea what was happening. I'd never seen anything like this before. It was so jarring. The man with the gun and his cohorts took all three of us back upstairs.

I put the key in the lock and opened our front door, fumbling and shaking the whole time.

Little by little, many more agents arrived. They separated Yegi, Reza, and me, sitting us on opposite sides of our apartment. Most of them wore surgical masks to hide their identities. Several had video cameras and were filming everything.

They ripped our home apart—ransacked our place. They went through everything. They even cut open tea bags. I don't know what they were looking for. Later, we were told, it was "microchips" they were after.

I didn't know what this was about, but I thought the only "evidence" that could potentially be used against us was a storage locker we kept full of liquor in our apartment's garage. We had a good stash: twenty bottles of hard liquor, ten or fifteen bottles of wine, and a hundred fifty bottles of beer. Two bottles or two hundred bottles, I knew it didn't really make a difference—it was the same crime. We had to lose the key.

"Booze. Flush," I said to Yegi, using words that, even if the agents spoke English, they probably wouldn't understand.

Yegi is smarter than I am. She figured it out, and she had the storage key on her key chain. She told the guard that she needed to use the toilet, and she was able to do it and tell me with her eyes that she had flushed the key.

When they asked us if we had storage, Yegi said, "No, the landlord uses the storage." This is common in Iran, and they accepted her white lie; also common.

We sighed with relief.

More people showed up, agents filling the seventeen hundred square feet of our two-bedroom apartment— the big living room with a terrace, the kitchen, the bedroom we'd converted into an office because we both worked from home. The agents walked over our hardwood floors covered with antique rugs.

They confiscated all of our electronic equipment and told us to put our cell phones and laptops on the dining room table.

"Give us all the codes for accessing your devices, email, and Facebook."

I thought, *I have absolutely nothing to hide.* So we did as we were told.

As more agents showed up, one female guard arrived in the full Islamic covering known as a chador, which means "tent" and is aptly named. She took Yegi into our bedroom and forced her to open our safe. The only things in it were a few thousand dollars and euros in cash, our identification and passports, Yegi's jewelry, her sealed U.S. immigration documents, and powers of attorney giving me the right to liquidate our family's Iranian assets.

After a torturous hour and a half, they walked us out of our apartment, put us back in the elevator, and took us outside through the garage. We lived in a neighborhood comprised of high-rise apartment buildings built by American and Israeli construction companies in the 1970s—the best-built buildings in Tehran—and known for being a more westernized, less outwardly religious neighborhood. It happened, also, to be in the west of the city.

The three of us were surrounded by several people that were clearly not from that part of town—several of them with surgical masks on and the one woman in full Islamic covering—and stood out at eight thirty on a weeknight in Ramadan. There were a lot of people coming and going. We had to walk by neighbors. People in Tehran have seen this kind of thing before. They know not to get involved.

We were led to an unmarked white van with tinted windows that also had curtains over them. They handed each of us a blindfold and told us to put it on. They took my glasses. I see very poorly without them; my vision gets blurry and doubled.

They cuffed my hands in front of me; they didn't put handcuffs on Yegi and Reza. The only other time I'd ever been handcuffed before was when I had been cast as an extra in a History Channel production in which I played a member of the Ba'ath Party in Iraq whom Saddam Hussein had executed.

"Don't talk," they told us.

A guard sat next to me and asked, "Are the cuffs too tight?"

I thought, *You see this in the movies all the time.* When they ask a question like that, if the suspect says yes, then the cop tightens them even more.

I said, "No, they're not that bad."

"No, no," he said, "if they're too tight, I can loosen them a little bit. This is not Guantánamo." And he took the key and loosened them, and I sat there with my hands loosely restrained. And I thought, *This can't be that big a deal.*

We drove for a short time. When we got out of the van, still cuffed and blindfolded, I told Yegi in English to get in touch with a fairly powerful distant relative of mine if she got out first. I assumed she'd be free within an hour. I thought, *We just need to nip this in the bud as soon as possible. If whoever I know can intervene, it will all blow over.* After that, we were separated.

2

Us/Our Life in Tehran

At that moment Yegi and I had been married just fifteen months. I was thirty-eight years old and there was still a lifetime of things my wife didn't know about me. I had habits and pet peeves; things I was afraid of, others I wanted to try. Funny stories that I hadn't told her yet. The sorts of discoveries that all couples make during their first years living together. But my having a secret identity as a high-level intelligence operative wasn't among these.

I knew she didn't believe I did, but I also knew that both of us had just been thrown into a situation that had immediately changed everything we had understood about our lives.

Before getting married we had dated for four years— no small feat in Iran, where it is technically against the law to have a relationship out of wedlock. During that time Yegi's eyes had been opened to many new ideas. Before even meeting me she had studied English translation and had earned a master's degree. Then she spent several years as an international news reporter. She traveled more than any middle-class Iranian twenty-something could imagine, getting as far as America, and came back again. But through all of that she lived at home with her parents—sharing a bedroom with her older sister—until the day we were married.

I had accepted the responsibility of providing her with a good life, although I always suspected that she might become a more successful breadwinner than I ever could. She was smarter, more ambitious, and younger. She had just turned thirty a few weeks before.

Here we were, at the outset of our life together, moving in all the right directions. Three days away from coming back to the U.S. to claim her green card so we could begin the bicontinental, bipolar life we envisioned for ourselves.

All of that was being taken from us. She already realized it. But I didn't.

It's been widely and erroneously reported that I moved to Iran in 2008 because of a love affair with my ancestral homeland. It's the sort of detail that seeps into an ongoing news story often enough that it becomes fact. In this case, though, I have the opportunity and responsibility to set the record straight.

Yes, it was out of attraction that I finally—after years of brushing up against it and later longer, drawn-out encounters—moved in with Iran. But this wasn't an unconditional thing. It was definitely not meant to be. I had to put in a lot of effort to make it work.

Iran was the slightly less polished, funnier sister with real depth who never dated much in school. Iran wasn't flashy and sexy on the surface. Iran's not France or Japan. But it's not like it's hard to love either. It's just *Iran*.

The spell it cast on me was slow to come on but ultimately very powerful. It can be a hard place—look at the dry and rocky landscape or Tehran's pollution—that is capable of incredible tenderness. Some of the softest people you'd ever want to meet. Don't be confused by reports from visitors who have nothing but love and admiration for Iran, though. They only went for a visit and never tried to accomplish anything there.

You don't get to know it all at once. It reveals itself in pieces and that, too, can be part of an elaborate scheme to win you over. Don't buy it. In the end Iran will disappoint you if you let it. At least that's what I would tell friends and visitors who seemed to be falling too hard.

And sometimes I had to tell myself, too.

It's fair to say that my relationship with Iran is complicated, but I didn't move there in 2008. It was 2009, and there's a big difference.

If part of what you empathized with me about was that I was locked up abroad in some faraway land, I have to set that straight now, too. I was imprisoned in what had become, over five years of living there, the city I called home.

It would never be my "hometown," because that's San Rafael, California, but Tehran was the place that, with all my faculties and the resources available to me, I chose, for better or worse. That's important. It was my choice. No one sent me there.

When the world financial markets were collapsing and I decided that sticking it out in the Persian rug business would lead to personal financial—and perhaps actual—suicide, there was no second thought about where I would go.

I had made up my mind that I would be starting fresh in Iran. I'd just turned thirty-three. It was my Jesus Year.

While a lot of my friends were starting to have kids, I'm sure most of them thought I was a little crazy. Although a few were definitely envious. "I'm going global," I told them.

I knew, though, that given the circumstances, mine and the world's, going to Iran to take a stab at a career as a correspondent made more sense than any of my other possible choices. It was literally the only thing for me to do. I saw it clearly, although the path was fraught with obstacles.

I had a unique service to provide that very few others could or would. It was my own niche market, and in true entrepreneurial spirit, I decided to take that love and turn it into a career. It was my version of that very American can-do attitude.

I went to Iran so you don't have to!

As with any new enterprise, the early days were lean, and staying afloat in an unfamiliar game became my main goal. There was absolutely no good reason other than my own lack of fear—or was it lack of awareness, perhaps?—that for nearly five years I would be the lone American reporting on a permanent basis from Tehran, but that's exactly what happened.

In that part of the world, where permission is required to work as a journalist, I was one in a growing trend of reporters with deep foreign roots doing a reverse migration, half-knowingly putting ourselves in very precarious, and at the same time very essential, positions. Increasingly, it's people like me who bring you the news from the world's furthest-flung places, especially the ones run by Muslim authoritarians.

Barack Obama had just been elected and moving to Iran felt to me to be the most American thing I could do.

The truth is that nothing was perfect for Yegi and me, although we could measure progress in our careers and in our future together.

After we'd been a couple for less than two years, in the spring of 2011 she was offered a job as Bloomberg's Tehran correspondent. This was an unprecedented achievement for a young native Iranian journalist and she rose to the occasion. For years in Tehran's government offices she was known simply as "Ms. Bloomberg."

In the spring of 2012 Yegi and I went to the U.S. for a visit. I wanted her to see America and meet my family. Our relationship was getting to the stage where there was a solid foundation in place and the building blocks were starting to take on a structure.

It was during the first days of that trip that I was contacted by the *Washington Post* and asked if I would be interested in their correspondent job in Tehran. In one of my interviews I was asked if I could put my opinions aside and write news. Of course I could. I never reported anything that wasn't accurate. How could I say no to the platform that the *Post* would provide?

One of my references, a DC-based Iran analyst who was a regular source, likes to recount the call he had with the editor who hired me.

"Why do you think Jason is the best person for the job?" asked the editor.

As my source tells it, he laughed and replied, "For a lot of reasons, Jason is the *only* person for the job."

And that was probably true. There was no one else available in Iran at that moment who could write for a major English-language publication and no one working outside that would have been granted a working visa for the *Washington Post*. I was a journeyman, but one who could complete a sentence and whom the authorities in Tehran already knew.

My first editor, Griff Witte, and I visited the press officer at Iran's permanent mission at the United Nations to formally request permission for me to be the *Post*'s Tehran correspondent. Over tea the bureaucrat said, "I don't always agree with Mr. Jason's work, but he is fair. And as Mr. Jason will tell you, Iran is a pluralistic society and we welcome rigorous debate."

Leaving that meeting Griff and I shared a laugh over the "pluralistic" line. But we walked away confident I would get that permission.

When Yegi and I returned to Tehran that spring it was to the beginnings of a new stage in our lives together. Within a few weeks I was filing stories for the *Post*. I had a real paycheck every month—the biggest one of my life—and could claim expenses.

More than that, Iranian officialdom treated me differently now, too, giving me access to cover more prominent events.

Suddenly Yegi and I, back from a reinvigorating break that solidified our relationship even further, made up a very big part of America's news coverage of Iran, she with Bloomberg and I with the *Post*. We took it as a heavy responsibility, but also an incredible opportunity.

I finally felt that I was living up to some of the life expectations I had set for myself. It had taken a while to get there, but this was it.

In November, feeling established and stabilized, I made plans to visit Yegi's parents. This was a technicality. By then they knew me, but in Iran's family-oriented and strictly guarded social hierarchy I was still an "outsider" or "no one" to them. Farsi, for all its ritualized politeness and reputation for being deeply poetic, can also be incredibly unsentimental when it wants to be. I knew that well.

Armed with sweets and flowers on a rainy November Tuesday night, I made the trip across town by taxi alone, without the traditional male chaperone who would accompany a prospective groom—a break from tradition. At thirty-six I felt, really for the first time in my life, like a grown man.

I arrived and presented the small gifts and we sat for tea. Everyone there—Yegi, her parents, and her older sister—knew what was happening, but no one let on that they did. Finally I asked if she and her father would join me in their sitting room. I wanted to discuss the future with them.

We sat down, he with a pad of yellow paper and pen, prepared for the occasion. We quickly entered into a more in-depth conversation about my finances than I've ever had with anyone—Yegi's dad is an accountant, but more than that he's an Iranian dad—and came to some agreements about what he expected of me as his youngest daughter's future provider. Once the parameters were set we both—he and I—signed. I'm not sure if it was a legally binding contract, but given what I now know of Iran's legal system, I'm guessing it could have been.

It sounds more uncomfortable than it actually was. Within twenty minutes Yegi and I were engaged.

As we planned our wedding for the spring of 2013, politics inevitably crept into our personal life. Early June would be the next presidential election in Iran and we needed to leave enough time to fit in a honeymoon before the campaign season heated up. That was my only condition.

Quietly there were concerns over the fate of Western journalists should a hardline candidate win. We were bracing ourselves for the possibility that our lives might take a drastic turn, doing all the preparations for Yegi's green card application so we would be ready to leave if we found ourselves out of work.

We made arrangements to hold the wedding in a "garden" on the outskirts of Tehran. People who want to have a mixed-gender wedding party with dancing and the opportunity for women to let their hair down have to seek out such venues.

There is a thriving industry of wedding planners who have access to sites where, for whatever reason— usually the potential for steep financial rewards— the owner is ready to take the legal risk of hosting an event where a multitude of Iranian laws will inevitably be broken. When we visited the property, which was inside a maze of industrial alleys, it was in the frost of late February and I was skeptical that it could be transformed into a fairy-tale wedding venue. But I was underestimating the power of spring.

As we planned our wedding several of my American friends were ready to fly to the other side of the world, despite their families' warnings, to take part in our nuptials. Yegi and I went about asking for permission to have American guests at our wedding in a completely transparent way.

"Perhaps after the election," we were told by the foreign ministry. "It is too sensitive a time. No one wants to take responsibility for American tourists at the wedding of an American correspondent to an Iranian one."

But we knew better than to be overly disappointed. Living with letdowns is a skill essential for surviving life in Iran.

On April 19, 2013, the show went on, minus the presence of many people who should have been involved and in front of a few who went to great lengths to be a part of our special day. My mom had arrived a week in advance to help with preparations. My big brother Ali came alone, leaving his family back in California. He trekked all the way from the other side of the world and stayed for less than two and a half days. "I wouldn't miss it, Jason," he told me; I knew that this was the only thing that would possibly bring him to Iran. "I'm happy to see you so happy," he said.

Our friends from Paris, Charley and Marie, and my old friend Mathias, Charley's cousin, all came, too.

The presence of the five of them was enough to make it real for me.

I'm not much of a dancer, but I didn't sit down that night. Yegi came home with me to the apartment we'd rented for the first time and we started our life.

A few weeks later Hassan Rouhani, a regime loyalist who was dubbed a "moderate" for his pragmatic promises to lift Iran out from under the weight of sanctions by engaging more diplomatically with the world, won a resounding election victory.

Instead of having to think about a possible exit strategy we were suddenly in the place everyone wanted to be. Iran announced it was open for all kinds of business. Everyone welcome. People were coming to me, and the ones I contacted were eager to connect. Suddenly an email with the subject "Greetings from Tehran!" was guaranteed a response.

During those months our home became a landing pad for many foreign visitors. Our apartment in Tehran's Shahrak e Gharb—literally "West Village"—was a large two-bedroom flat with a balcony on the third floor of a pre-revolutionary luxury high-rise, making it a sought-after address for people in the know.

Unlike the majority of nonnative expats we paid our rent in rials, the local currency, one of the many perks of marrying a local. Most foreigners, whether they were diplomats, journalists, or executives of large companies, were forced into paying exorbitant rents in dollars. In cash. Our rent for a seventeen-hundred-square-foot flat was the equivalent of about $800 per month.

We took regular international trips, and when you're in the Middle East a lot of the world is not that far away. Thailand, Sri Lanka, the Maldives, Istanbul, Paris. And on nearly every one of them we'd work in a layover of a couple of days in Dubai—I had earned good frequent-flier status on Emirates—to shop for all the items that sanctions made it hard, or too expensive, to find in Tehran. Halogen lightbulbs, coffee, cheese and chocolate, high-quality underpants, and Christmas ornaments.

I hired a woman who also cooked and cleaned for another journalist friend. She came to our place twice a week for six hours, on Sundays cooking for the rest of the week and on Tuesdays to clean. I paid her the equivalent of about twenty dollars per day. My shirts were always ironed for the first time in my life, so Yegi was happy, too. Locals said I paid her too much. European diplomats I knew paid their help sixty euros a day, and theirs didn't cook.

I never argued with the naysayers, though. If you're a foreigner in Iran local people will always tell you they can get you a better deal. In my experience listening to such "friends" invariably had hidden costs. To the extent that I could, I tried to cut out the middleman, unless I could figure out a way to be the middleman, which, in Iran's middleman economy, is who you want to be.

Ours was one of the most classically Persian-decorated homes I'd been to in Iran, putting it in stark contrast with the majority of places we'd go. Most urban Iranians have, for reasons beyond me, shunned everything natively available for Chinese rip-offs of what I imagine went in the homes of gaudy nineteenth-century low-level French nobility. Lots of uncomfortable-looking chairs. And crystal.

What do you get the Iranian bride that's got everything? A crystal donkey.

I had seen the crystal version of just about anything you can imagine.

But our place was different. The parquet floors were covered with antique Persian rugs I had repatriated from my family's California supply. We found a local craftsman who lovingly refurbished old wooden chests and tables, touching them up with classical-style miniature paintings. Some of our serving trays dated back to the nineteenth-century Qajar era.

On the surface I ran the risk of being confused for a modern-day Persian T. S. Eliot type, trying tragically too hard (and unsuccessfully) to go native. People who know us, though, will vouch for the fact that we weren't going for pretension. We just wanted our home, for guest and occupant alike, to be a place where people felt comfortable and welcome.

For us it was a sanctuary. Against all sorts of odds, my life was stable and where I wanted it to be. And that happened to be in Tehran.

I ventured out into the city less and less. I almost stopped going to press conferences entirely. Why should I be bothered? They were played live on television, and even when I made the long trek into the seat of government offices, far down in old Tehran, officials almost never allowed a foreign journalist to ask a question.

I saved my energy to do real reporting on stories and people that required some depth and connection.

Pollution in the west of the city was thankfully less than in the heart of Tehran's mayhem. Our building had twenty-four-hour security, at the gate and the front desk; parking; and underground storage space for each apartment. We used ours to store our ample liquor supply, collected from diplomat friends who generously understood the restrictions the rest of us had to live with.

We were living a life that others envied. Yegi and I both had jobs that paid us in dollars, putting us in a high earning bracket. Life was good.

I was being led, blindfolded, through corridors and finally into an air-conditioned room. At the door I was instructed to take my shoes off—in Iran, it is customary to take one's shoes off indoors. Two men, whom I couldn't see, sat me down in a vinyl chair.

There were a lot of other people in the room. I could hear whispers, and people pacing, and prayer beads being thumbed. I could smell the competing body odors of different men.

After a few minutes, a male voice addressed me.

"Do you know why you are here, Mr. Jason?"

"No," I said, turning my head in the direction of his voice.

"You're the head of the American CIA station in Tehran," he said. He never raised his voice, but he was accusatory: "We know it. And you have a choice. Tell us everything, and you'll go home. You'll get on that flight to the United States on Friday as planned, but you'll be starting a new life working for the Ministry of Intelligence of the Islamic Republic." The offer was absurd in its directness and so I didn't think he was completely serious.

"If not, you must change your clothes. When you put the prison clothes on it's not clear how long you're going to be here. The odds are you will spend the rest of your life as our guest. You'll never get out of here. So tell us everything."

"There's nothing to tell," I said. "I'm just a journalist. You've made a mistake. This is all wrong. I'm just a journalist."

"*Just a journalist* has no value to me," said the voice.

I was trying to rationalize with somebody whose logic was very different from my own. He had his position and he wasn't deviating from it.

The voice started throwing out names of well-known Iranians and Americans, people I knew, people from the news, and people I'd never heard of. "What's your relationship with John Kerry? What's your relationship with Obama?"

"I've never met either of them," I said, which was true. The idea that I knew the top officials of the United States was ridiculous. Everything I said, though, just seemed to make the hole I was in deeper.

I tried to talk my way out of it. I explained to him that the work that I did was for the *Washington Post*. I explained that I was permitted to work in this country. I said this was just a misunderstanding. I told him to call the press ministry. They had literally just reissued a one-year extension of my press credential *that morning* for Christ's sake.

"You're a spy. We have all of the proof. And you just need to tell us," the voice said calmly. "Everything."

"If you have proof, why do I need to tell you everything?" I asked.

"We need to know that you're reliable. That we can trust you to cooperate," he told me.

"I'm not reliable," I said. "I don't work for America and I'm not going to work for you. I work for the *Washington Post*."

There was a pause, and he began talking again as if he were reading from a secret memo.

"'Alan Eyre,'" the voice said. "'Avocado. T-shirt.' What does it mean?"

I thought, *Okay, I can explain all of this.*

Alan Eyre was a diplomat who happened to be the State Department's highest-ranking Farsi speaker, and for that reason alone the Iranian regime regarded him with suspicion. He had been based in Dubai for years, which is where I first met him. I had just run into him the week before at the nuclear talks in Vienna.

In 2010 I'd launched a project on Kickstarter, the crowdfunding website, about why there were no avocados grown in Iran. Of all of the many things I had seen in Iran over the years, the most troubling was one thing that I *didn't* see. There were no avocados to be had inside the Islamic Republic. So at a time when it was too risky to cover the day-to-day politics, Iran's lack of avocados became an obsession I had to get to the bottom of.

Where was the guacamole?

The Iranian Avocado Quest was an attempt to make a point. The very simple fact that the beloved avocado was almost unknown in Iran proved the first part of my argument: that Iran was cut off from the world, even in benign ways. The project would help explain part of the issue: what stands in the way of building a bridge, even a seemingly frivolous one.

Many folks took it as a joke, but that's sometimes the best way to get people thinking about a new subject. I ended my project pitch with a plea:

"I think the time is now for the American people to connect more closely to Iranian society however they can. And I'm offering a bridge to do just that. Hope you join me for the ride. I will bring the avocado to Iran, but I can't do it without your support. The future of Persian guacamole is in your hands."

In return for each $20 pledge to help me start an avocado farm, if successful in fund-raising, I would distribute T-shirts to funders. Including Alan Eyre. The thought of my avocado project made him laugh and he'd wanted to pitch in a few bucks when I had seen him once in Dubai. He was, he said, in it for the T-shirt. It didn't matter; the project failed to reach its funding goal.

"It was a joke project that failed," I said from beneath the blindfold. "There's been a misunderstanding."

"No, there's no mistake, Mr. Jason. It means," the voice answered his own question, "you are the head of CIA operations in Iran. This is our proof that you are what we say you are."

"This is ridiculous." I was still trying to make light of the situation. Something I had done a million times before in my life and it almost always worked.

"Perhaps, but if you're *just a journalist,* why would you have contact with Alan Eyre?"

"I interview and sometimes communicate with people, including officials. This is a normal part of the job."

I tried to further explain Kickstarter and the significance—or lack thereof—of avocados, but blindfolded and under duress in Farsi, and speaking to an unknown audience, I had reached the limit of my capabilities. This went on for a few minutes.

They brought Yegi into the room. She was crying, obviously struggling.

"Jason, what's going on?" she asked. "They've changed me into prison clothes. Why are you not in prison clothes?"

"This is all going to finish soon, baby," I told her. "Just be calm."

"Have you done anything wrong? They're saying terrible things about you," she said clearly but tearfully. "Jason, just tell me you're not a spy."

"Of course I'm not a spy," I said.

"I know," she said. "I love you." She wasn't in the room for more than two minutes. And then they took her away, and there was no more sound.

"Mr. Jason, you still have the opportunity to tell us everything right now," the voice said.

Another voice interjected; this one had the distinct and musical accent typical of people from the city of Esfahan. "Dear Jason, the Great Judge is making you an offer and he never breaks his promises. Just tell us what you know," he said.

"I have nothing to tell you," I said.

"He's afraid," the first voice, the one that belonged to the Great Judge, said to the others in the room. "He's afraid. Change his clothes and take him to the cell. He'll start talking within a month."

I'll start talking in a month? That's pretty over-the-top, I thought. *There's no way I'm going to be here more than tonight, and maybe tomorrow. Their job is to scare me.*

I had too much working in my favor. The press ministry would be on my side. The foreign ministry would be on my side. President Rouhani was in the middle of negotiations on the nuclear talks that he needed to work. The *Washington Post* would be on my side.

But it was the first real moment where I thought, *This might be worse than I think.*

They led me out of the room where I'd spent the last half an hour being interrogated and down an outdoor corridor into another room, where they took off the blindfold and the handcuffs. It was even more confusing without my glasses.

We were in a small infirmary. There was a patient's bed built into the wall. I saw cotton swabs and tongue depressors, a blood pressure cuff, and the thing with the little bulb a doctor uses to look into ears. There was someone there in a white coat—they called him "Doctor," but who knows what he was.

They told me to take off all my clothes except my underwear.

They weighed me. They took my blood pressure. I was shaking.

They handed me a set of their version of prison blues. Light blue pajamas, basically. Pants with an elastic waist without a clear front or back, and a shirt with four big plastic buttons. They gave me a pair of flip-flops and a pair of prison underwear—darker blue.

Then they led me through a hallway and stopped at a door. They pointed in, and that was that. By the time I actually got into the cell on the night I was arrested, it was past midnight.

The cell was small, about eight and a half feet by four and a half feet; I could lie down completely in one direction but not the other. The ceilings were ten feet high. There were two windows above, with bars on them that let in light but no view. Those windows let me know approximately what time of day it was.

An aluminum door led to a toilet—a hole in the ground, as they are in that part of the world—and a tiny sink. The door had many things crudely engraved in it. None of it was in English. There were many rows of four lines with one line cutting through. Exactly like you imagine from the prison movies you've seen. Some of those sets of lines added up to more than a hundred.

There were two blankets and a dirty, crudely cut fragment of a machine-made Persian rug, in a Kashan pattern, with an elaborate floral design, on the concrete floor. That's what I would sleep on, or try to.

There were two fluorescent bulbs on the ceiling that I quickly learned would stay on twenty-four hours a day. There was a fan in the room that made a crazy amount of noise. They obviously didn't want me to actually sleep. It was extremely warm.

The door had two holes: one at eye level that they could open and talk to me through, and one down below, which, I figured out, was for food.

There was a copy of the Koran in Farsi and Arabic.

I don't think that first night I slept at all. I was just waiting for somebody to come whom I could talk some sense into.

I sat, but pretty soon I started to pace. I ended up spending a lot of time in that cell walking back and forth. I was so confused, and so unprepared, but still optimistic that this thing, whatever it was, was going to go away. I would be out in a couple of days. *I have things working in my favor.* But I was very afraid for Yegi—where she was, what they might be doing to her, what she was going through.

I thought, *I can handle this. I have to handle this.* But I had a real feeling of responsibility for my dear young wife who had never been put in any sort of perilous situation before. Neither had I, actually.

The call to prayer always starts whenever the first sliver of sun comes up. Around four A.M., I heard it. During Ramadan, that's when people eat, before they begin their fasting for the day.

A guard I couldn't see brought me water and some food: herbs, a piece of cheese, a couple of walnut halves, and a single sheet of bread called lavash.

In those lonely hours I thought about my dad, and our migrations in reverse: his to America as a young man, and mine back to his homeland.

We grew up on opposite sides of the world. Dad was born in Iran's holiest city, Mashhad—or the "place of martyrdom"—home to the shrine of Shia Islam's eighth saint, Reza, the only one of Prophet Mohammad's descendants at the heart of Islam's early split who's buried in Iran.

Mashhad was then and is now an important city. An epicenter of Islam and a crossroads of people and cultures.

My grandfather Hajj Kazem Rezaian was an influential person in Mashhad. He was a stakeholder in a range of local businesses including a nearby turquoise mine. He also, as his forefathers had before him, held a key role overseeing the Imam Reza shrine complex, which today is one of the world's largest charitable organizations. He and my grandmother married young and wasted no time in having a family. They had nine children who survived infancy and lived to adulthood.

My dad, Taghi, was the third of those. As fate would have it he was the second son. That, in Iranian cultural terms, meant a life destined to be lived in the shadows. But my dad wasn't really a shadow sort of guy.

He had a car by the time he was seventeen, unheard of in Iran at that time. Although Tehran was filled with bars and nightclubs, catering to thousands of foreign workers and a growing westernized middle class, Mashhad still had a ban on what the Shia seminary still deems excessive: dancing, most live music, drinking, and commingling of nonmarried members of the opposite sex.

My dad was not a drinker, but I just can't imagine him in a place where he wasn't encouraged to show his feathers.

In 1959 Dad decided he wanted to leave Mashhad and continue his studies in America. This was an idea that had no precedent in the Rezaian family, but my grandfather was a forward-thinking man of significant means.

He supported his son's dreams and tapped his connections at the U.S. consulate, who helped my dad, a strong math student, get admitted to Georgetown.

There was one last obstacle: my paternal grandmother, who didn't want to let young Taghi go. Of her nine kids, my dad was one of her favorites.

"It will break my heart if he leaves, because he will never return," she cried.

"But if we don't let him go do you think he will ever forgive us for standing in the way of his dream?" Kazem asked rhetorically. "He is going. Let him go."

With that Dad took off. First a train to Tehran, and then one to Baghdad. A bus to Beirut. Then a flight to Rome. Another one to London. And a final leg to Washington, DC.

Years later he would laugh about how far away he was in the beginning, only accessible by mail that took weeks to arrive. And then how easy it became to fly Iran Air direct from JFK to Tehran, only to have that route disrupted, along with so much else, in 1979.

He was miserable in DC from the outset. It wasn't the America he knew from the movies. His roommate, the only other Iranian at Georgetown, was fanatical in his adherence to Islamic rules, chastising my dad for not fasting during Ramadan. Dad was indignant. He knew the rules; he lived in a place steeped completely in the Koran, which clearly states that when traveling or far from home fasting isn't required. He was living in DC, yes, but it was not home.

Who the hell was this guy to tell him how to practice his faith?

Well, he was Sadegh Ghotbzadeh, a young radical who, two decades later, would become a top aide to Ayatollah Khomeini and the Islamic Republic's first foreign minister. He would become involved in secret negotiations over American hostages taken at the U.S. embassy in Tehran.

He would also go on to be executed in front of a firing squad for supposedly plotting to assassinate Khomeini.

But this was before all that. My dad knew he couldn't live with the guy any longer and he decided he was going home to Mashhad even if it meant he had to fast during Ramadan again. He was willing to admit that he had made a mistake. America wasn't for him. He put it all in a letter to my grandfather, expecting Hajj Kazem to tell him to come home.

The reply took several weeks, and it wasn't at all what Dad expected.

"You will not be coming home. I have sacrificed a great deal to send you to America and you will get your education. If you decide that you want to return after you are finished, Mashhad will be here. Wait for instructions in my next letter."

Weeks passed before the next letter came. In it my grandfather described another "university" where several other Mashhadis were studying. He suggested that my dad go there, where he would be welcomed by a budding community of fellow Iranian students.

He packed his belongings, took the letter to the bus terminal, and showed it to the ticket seller.

"Send me there." He got on the bus. He was twenty years old.

I can't know for certain, but it's a safe bet that my dad is the only person in history to leave Georgetown, knowingly and willingly, for Napa Junior College.

In Northern California he and his new band of Iranian transplants felt comfortable. It was a small manageable town with ample work. They were learning English and the community accepted the newcomers. Why wouldn't it? They were all young men from well-to-do families from a country that was not only one of Washington's closest allies but was also considered exotic and mysterious. There was no Islamophobia yet. No one in America had ever heard of an ayatollah.

Dad excelled there because he hustled.

On the weekends and over the summer, he and a friend would go to Lake Tahoe, where they found work as busboys. Two of them did the job of four in exchange for the pay of three. It was a good life, and slowly his urge to return to Mashhad began to diminish.

It was 1963 and he had completed Napa's associate's degree program. He and his crew moved to the Bay Area, some of them opting for Berkeley and the rest, including him, headed for San Francisco State.

He couldn't have imagined then all that was to come for the Rezaians in the next half century.

3

A New Way to Look at Iran

During the months that followed Rouhani's un-expected June 2013 election there was a semi-softening around the edges. A steadily increasing number of foreign journalists and their camera crews began coming back to Iran. University tour groups, the kind that charge exorbitant rates for alumni to travel with supposed experts, started showing up again after a long absence.

And so, too, did the outliers. Backpackers, hippies, and what I think of as specialized tourists, those people in search of unique experiences. Out-of-bounds snow-boarders, ravers tapping the underground electronic music scene, fine-art speculators, and people trying to cozy up to Tehran's LGBTQ communities.

Some of these quests are more obnoxious than others, but the common thread was that they were all part of a process of discovery. The unmistakable reality that Iran was "hot." The latest *it* destination, which made every travel magazine's "must visit" list in 2014.

Everything I had been saying for years about Iran's being a traveler's paradise was being recognized. Two years earlier I'd written articles in the *Washington Post* about tourism in Iran—its potential and the challenges it faced—and I was laughed at. Now, seemingly, the whole world was lining up to come to Iran.

I'll be honest, I felt some satisfaction. Of all the story lines that didn't fit within the old Iran narrative, I'd helped take one of the hardest-to-sell ideas and turned it into a trending story.

Zarif and Rouhani believed they'd made it happen. In fact there were so many other factors, and one of them was me.

In the early months of Rouhani's presidency it became clear that the image of Iran was evolving. After the Ahmadinejad years foreigners, including journalists, not only felt safe again but were being wooed.

By the spring of 2014 I started to alter course, in my mind at least.

For anyone who wanted to accept things as they were, it was becoming obvious that Rouhani and the

U.S., under Barack Obama, were trying to find points of understanding. Besides the nuclear negotiations there was the emergence of the Islamic State and the reality that Washington and Tehran's interests coalesced in a way that wasn't completely comfortable for either.

What I saw in front of me was a changing story, one that I could tell. At that moment, in the spring and early summer of 2014, I had the best of all worlds: permission to cover Iran from the ground and people in Washington willing to talk with me about it.

But I was also getting bored.

"I have to see what happens next," is what had sustained me through the dark times between late 2009 and Rouhani's election four years later. I was a newlywed and we were a couple with ambitions. It's a natural progression for anyone, I suppose, although our variables may have been unusual.

We were angling for a life on two shores. Others were doing it, and it seemed to us that we were in as good a position as anyone to make it work.

Change, as always, was slow, but we knew it was coming.

One of the side activities that kept me engaged was introducing actual people, not just through my reporting, to Iran. Whether I helped plan a visit or just dropped in for a meal or a museum tour, it was part of my joy. Meeting Americans in Tehran was particularly fulfilling. So I started collecting those experiences.

At first it was a trickle. In the years since the revolution an average of fewer than five hundred non-Iranian Americans had visited Iran annually. And that included journalists and aid workers. But the numbers started to rise and in 2013, following Rouhani's election, the floodgates opened in a way they never had before and admitted a wave of American visitors.

As the lone American citizen reporting on a permanent basis from Iran I was a natural person to seek out. It didn't hurt that anyone who read one of my articles could easily contact me through the link at the bottom to my email address.

In 2014 there was a new group every week.

I met with several senior citizens who had been Peace Corps volunteers in Iran in the 1960s. I attended a lecture about organics by a blueberry farmer from Minnesota at Tehran's chamber of commerce, and interviewed him afterward. I gave talks to tour groups comprised of World Affairs Council members. And I dined with a group of American and Iranian business leaders who could sense the impending opportunities.

Some of these encounters I wrote about and others just sat in my mind, confirming what I already knew to be the case: like it or not, an opening was taking place, and as is often the case, it was normal people—intrepid ones to be sure—leading the way.

But I wasn't overly optimistic. Doors open and sometimes they close again. I had heard of too many Americans who were denied entry into Iran—like my friends who wanted to come to our wedding—and many more Iranians who, for no good reason, were unable to visit the U.S.

For me there were two things that became clear, though: Iran was not at all ready for the coming influx of tourists, and it was time for me to come up with a new and incongruous subject to write about.

I started thinking about food.

Whenever I visit a new place my research is never about the main sites or the climate I'm about to step into, but about what I'm going to eat when I'm there. I have chosen not to go to countries because I was convinced their food would suck.

With Iranian food there was so much to ponder. It was the ultimate expression of the country's identity: varied, resource rich, uniquely accented, sometimes pungent, hard to translate, and often unsightly.

Persian food can be gorgeous and fragrant.

But it simply doesn't show well the way Thai, Japanese, or Italian food does. One couldn't make "Visit Iran" posters featuring our most popular dishes and expect to entice people to come halfway around the world to try them.

Furthermore, Iran doesn't have the rich restaurant culture that other countries do. Most eateries lacked imagination. None of the love and hospitality that goes into preparing a meal for guests to Iranian homes existed in Iran's restaurants.

But suddenly there were people even coming for the food. I helped tailor two trips for food writers, one a half-Iranian American, like me, who had recently published a cookbook on Iranian cuisine, and the other a Dubai food blogger who led specialized eating tours of the Emirates' many ethnic enclaves. We discussed the idea of doing the same in Iran. And as I've been doing all my life, I connected the two of them based on their common interests, knowing that there is strength in numbers.

Within days of those two visits something remarkable happened: I was contacted by a producer from Anthony Bourdain's show *No Reservations*. They would be coming to do an episode in Iran soon and wanted to talk to Yegi and me.

At some moment, or maybe it was just how I was wired, I recognized that food was one of the best ways into knowing a place and its people.

In the case of Iran, as with every other matter, it was complicated. Iranians love to eat; perhaps the only two things they love more are shopping and feeding others.

The ancient tradition of stuffing guests became increasingly difficult for most Iranians during the extreme sanctions of the early Obama years, because people were busy figuring out how to keep up feeding themselves.

Iran wasn't near a famine or shortages, but stomachs weren't as full as they had been and that's usually a bad sign for a country's leadership. I used to joke that for many Iranians going to bed hungry meant they were only able to afford one skewer of minced lamb kebab in place of the ubiquitous two served in Iranian restaurants the world over.

No matter what the occasion, as is the case almost anywhere, food and the rituals around it are an essential component. Unlocking that knowledge can lead to deeper understanding.

But when I first started going to Iran, food and the other great elements about actually being in the country were my little secret, which always blew my mind. It wasn't as though Iran was an unknown country. It had had a very defined sense of itself for over 2,500 years when I finally arrived in 2001. And it's not small. At 80 million inhabitants, it's among the larger populations on earth. Couple that with a well-documented brain drain that has led to the flight of millions of Iranians in recent decades and one might wonder why there aren't more Persian restaurants in the world's great cities.

I call it my secret, but I was, from my earliest inter-actions, doing everything in my power to uncover Iran by covering Iran. Later this would be spun into a sin-ister narrative that had me exposing elements of public life in an attempt to find an Achilles' heel. Iran was the Death Star and I was R2-D2.

To tell you the truth, though, I wasn't getting much traction. Before moving to Iran I had been a fixer for many reporters. And more than that, I'd consulted with producers and correspondents ahead of their reporting trips to Iran.

In 2007, at thirty-one, with no consistent work and struck by Anthony Bourdain's curiosity, I did some-thing I did often those years: I sent a cold email, not expecting a response. In it I suggested that the show's producers and Bourdain consider doing an episode in Iran, and I gave a list of reasons why—some cultural and culinary highlights, a weird food or two to try, be-cause he still did that on every episode then—and why I was the guy, *the only guy,* to help them do it.

Weeks later I got a response. "Not only is Tony interested in doing an Iran show," an associate producer wrote me, "he has been for years." They wanted to know more.

Over the next few months in phone calls and emails I plotted a shoot with the associate producer, what it would look like, where we might go, how to get permissions. I had had plenty of these discussions before and knew that few of them actually panned out. Talking about a project and getting paid to plan one, I learned, are two very different things.

Finally I got a call from the show's executive producers at Zero Point Zero, *No Reservations'* production company, who wanted to meet me.

I headed to New York to visit their Tribeca offices and we talked about what might be possible and what definitely wasn't. They were interested and enthusiastic but not overly optimistic. I couldn't know if that was because of me or because of Iran, but I guessed it was the latter. The conversation continued moving in the right direction, until I got news that the network's insurance company wouldn't provide coverage for an Iran shoot.

For years afterward I'd get periodic messages from that original producer about his desire to come to Iran. A curiosity I helped pique.

But I put my foodie dreams on hold. Until the Rouhani election, when everything that I had waited for a more opportune time to pursue seemed feasible. In the spring of 2014 there was a buffet of Iran culinary journalism in the works. I just didn't realize that within weeks it would be my goose that was cooked.

That Iran has one of the most developed culinary traditions in the world and a people who can be violently hospitable—"a murderous generosity," Bourdain called it—but lacks a decent restaurant culture is one more of the incongruities that makes Iran so endearingly hard to understand.

But juxtapose this against what one finds if they're lucky enough to end up in an Iranian home. Dishes lovingly prepared for hours from unwritten recipes, passed down from grandmother to granddaughter to daughter, and repeat. Tables with an overflowing abundance of colorful rice—no one eats more rice than Iranians, except maybe Sri Lankans—each one with a different *tahdig,* literally "bottom of the pot," a piece of bread, or a layer of potatoes, or sometimes as simple as some rice infused with yogurt and saffron, that becomes golden brown under the heavenly mound of buttered rice, protecting it from burning, simultaneously becoming the most fought-over item of the spread when the pot is flipped over and magically appears.

At that time I was thinking a lot about the power of food and of sharing it. I was laying the groundwork for a possible pivot away from news, which I still loved. I was hopeful, though, that the biggest news issues propping up America's long enmity with Iran were getting resolved. I envisioned a diminishing need for the sort of coverage newspapers traditionally provide. The status quo on Iran reporting was changing.

As I thought about what would come next, I never thought about moving to report from another country, but in all honesty I was also getting tired of being in Iran all the time.

Yegi and I talked often about what we wanted from the future.

Running specialized tours of Iran, especially culinary ones, seemed like a great recipe for cultivating the life we wanted. If nothing else was certain, Iran's popularity as a destination among global travelers was on the rise and what they wanted from those experiences was one more element of the Iran-U.S. problem that I understood better than anyone else.

People wanted to touch, feel, and taste Iran. Very few of the local tour operators were in that game. Either they were too connected with the state security apparatus to veer from the prescribed and tired itineraries or they were too afraid of it. Regardless of the reason, when people visited Iran, if they passed through my hands, for a lunch or a tea or a trip to the rug bazaar, they invariably walked away a little fuller than the other guy.

I could see the brochures in my mind: "Visit Iran. It's not *that* bad." They read.

As I pieced together the previous several years, there was nothing that warranted my being thrown in solitary, and yet there I was.

From the earliest moments in my cell it seemed like a joke. The culmination of all the bad prison movies I'd ever seen. The disorientation of not being able to open the door yourself sets in quickly.

Remember the time you were stuck in an elevator for a couple of minutes or locked out of the house until the locksmith could get there? That feeling of weakness and fear about what will happen? It's like that for hours on end. Then days. Then weeks.

In solitary nothing happens. Well, almost nothing.

But then the notches you carve on the wall, in rows of four with diagonal lines through them, begin to add up. It's impossible to wrap your head around: three, seven, nineteen, thirty-six . . .

But you find mechanisms to cope. And if you're lucky you learn to quiet your mind, just a little, and live softly. It's not really submission. Don't do that. It's closer to an acceptance. You're being carried down a river and your odds of survival do not increase if you try to swim upstream. To the extent that it's possible, just go with it.

In solitary all manner of experience gets relived, because you are close to death. Not physically, but in terms of the very limited options available to you. Not hellacious, more like purgatory. It becomes a time to take stock.

I want kids, my mind kept telling me, starting sometime on the second day, which was strange since I had spent the entirety of my first thirty-eight years resisting the idea of procreation.

In the quiet of solitary thoughts can form. Ideas that take hold. And these can come in extremes: very real-feeling paranoia about the fate you might suffer, or delusions of grandeur about all the wonderful things that will happen if and when you ever get out. I did my best to avoid both and instead tried to plan—realistically—for the life I tried to convince myself would follow my imaginary release.

Having children, something Yegi and I had decided we definitely did *not* plan on doing, was suddenly front and center for me. I thought about *her* a lot, because whenever I imagine parenthood it's always as the daddy of an über-intelligent, extremely kind, witty, and cute little girl.

It was through this latest life plan that I began to know Iran, as a home, was over for us. There was no way I was going to be able to grow a family with Yegi in a society that is segregated along gender lines and systematically intolerant in so many other ways.

Soon I drifted to other thoughts, but this one always came back in my darkest hours. It was the promise of goodness in a murky future. It wouldn't take me long, though, to wander to things far beyond my control. I felt incredible guilt over the combination of worry and helplessness my mom and brother must have been going through, *again*.

They don't deserve this, I thought.

And I wondered how my in-laws would be coping with the situation. They had lived their entire lives in Iran. They knew the score. As the parents of three girls, one of them severely disabled and requiring constant care, they were the type of people who did whatever they could to stay out of trouble. All of their fears about their youngest daughter's marrying a foreigner, and a journalist at that, were coming true. Or at least that's what my mind told me.

All I really wanted to do, though, was see my wife. Hold her. And tell her I wanted to have kids with her.

Most of the time I just searched frantically for a clue as to why this was happening. I thought about everything that I had done up until that point, scanning for anything that could have led to our arrest.

The first time I met him, I could only hear his voice. I was blindfolded, which was the rule whenever I was not in my cell.

He spoke better English than I expected anyone there would and had a deep, breathy voice which immediately reminded me (and always will) of Wanda, the gender-bending character Jamie Foxx played on *In Living Color*—which consisted primarily of his wearing a blond wig and lipstick—who was constantly threatening to "rock your world."

In our initial encounter, before all the questions began, he claimed that he was chosen by the judiciary to defend me. "I am your attorney, chosen by the Great Judge," he said. I couldn't see him, but that was obviously a lie.

"If you're my attorney, why I am blindfolded?"

"It is for your protection."

"What does that mean?"

"The charges against you are very serious and you must tell me everything if you want to leave this place."

"What charges?"

"'Spionage," he said. "You are an espy. We know that. You will leave as soon as you like. Everything is up to you."

"I would like to leave now."

He was not actually my lawyer, he explained, but rather my interrogator, or as he referred to himself, my "expert."

"You must tell me about the avocado. This is code, we know that, but for what?"

If this is really about a Kickstarter project, these guys are dumber, are more paranoid, and have fewer real security problems than I ever thought possible.

"I'm sorry for saying so, but you're just making a mistake," I told him.

"No!" he said, slamming something against the wall close behind my head. My shoulders tensed and never really loosened. "We know it, and it will be much better for you if you tell us yourself than if we discover it."

From his voice, I assumed Kazem was a big guy, powerful. In time he would go on to become the perpetual good cop in group interrogations. But that voice. It made him sound so intimidating.

"We must execute you, Jason. You don't give us any choice. We prefer to let you live, but you refuse to cooperate."

On the fourth day, though, in yet another dead-end interrogation, fear and desperation mounting, I had a flash of inspiration. Something about our initial rapport made me think he might be responsive to affection, so I gave it a shot. Actually, I said . . .

"Right now you're my only friend in the world."

"I'm your friend?" he asked. "Really?"

"Yeah."

There was a long pause. I had no idea what was coming.

"Take off your blindfold and turn around."

I did what I was told to do, cautiously, and faced him for the first time. It was just the two of us in a sparse room. There were two plastic chairs and a single table against the wall I had just been facing. He quickly re-arranged the furniture so that we sat across from each other.

"Friends must be able to sit and talk comfortably," he said, but I wasn't convinced.

He was slim, wore square steel-framed glasses. His hairline was receding and had the three-day stubble that is common among those who would have you believe they are believers. He was probably in his early thirties. Very typical working-class Iranian. My nemesis was hardly the goon I had been imagining.

The initial interrogation sessions were awkward, in part because I didn't know the rules yet, the main one being that I didn't have the right to ask any questions. A concept I would be reminded of every time I asked anything, whether it was about my case or if I could use the toilet.

Despite Kazem's constant assurances that he was little more than a lowly cop doing his job and by no means a "top banana"—he actually said that, in English . . . often—I became convinced that he was indeed all-powerful. That he was calling the shots.

As frustrating as the circular sessions in those first days were, the hierarchy of the system holding me was not a concept I could even ponder. I was too busy thinking about the imaginary forces that must have been working so diligently to right this wrong, explaining to whoever needed to hear that this was a simple mistake. Family, friends, friends of the family, the press ministry, the *Washington Post,* the State Department. *It's just a waiting game,* I thought.

But they get to work breaking you down from the very start. As with the carcass of a hunted creature, it's easier to deal with pieces than the whole animal.

"No one is coming for you. The world now believes that you and your wife died in a car accident," Kazem told me.

Well, that sounds plausible. The number of road deaths each year in Iran is astronomical. I've reported on it, I thought, but said, "No one will believe that."

"Why not? You've been missing a week already and the story is over. Perhaps your mother is upset, but the U.S. government has said nothing. And neither has the *Washington Post,* but that is not surprising, since we know you never worked for them."

I tried to hold my ground, but I was slipping.

After a couple of weeks I started to bend.

Interrogations are never long enough, I thought, realizing that holding such a view, even for a moment, was not a good sign.

A guard would open the door, calling out, "Sixty-two, get ready." My prison code was 93-0-62, or just "Sixty-two" for short.

That was my cue to rise and put my blindfold on. During the earliest days there was a mix of anxiety and anticipation. *Maybe it's over,* I'd tell myself.

But, no, it was just getting started.

A guard would lead me down a short corridor, and then out the door that opened onto our walled walking yard, and then out another door to an adjoining building.

He would have a sheet of paper that he and the interrogator would both sign as proof that all official procedures had been followed. "See," sometimes they would tell me, "everything is recorded. This is all legal." I had no way of proving otherwise. It had been made clear to me in word and deed that I had no rights, especially not the right to question what was happening.

Once in the interrogator's possession I would be taken by the arm forcefully, but never painfully so, and led to one of the rooms. Kazem would sometimes put a hand on my midback and tell me, "Don't slouch. Stand up straight." Not in an intimidating way, but sort of like advice. It was as though he wanted me, as his adversary, to be tougher than I actually was.

Most of the rooms were big and open with plenty of space, but a few were tight, with mirrored divides for special sessions with "top bananas." In those instances the blindfold always stayed on.

Usually it was just Kazem, but on some days there would be another guy, Maziar they called him, who was obviously obese, which I could hear through his difficulties breathing.

"I will tell Maziar you said he was fat," Kazem joked. "He won't like that, but it's true. When we have pizza everyone has a half of one, but Maziar eats two by himself." I tried to feign a sense of camaraderie by giggling along.

But that's a hard thing to fake, especially as Maziar's primary role, as I came to understand it, was to threaten me with death or dismemberment.

To be fair though, there were carrots, too. "Tell us what we want to hear and your miserable life might be spared," Maziar promised as I sat facing the wall, told to give a full accounting of my experience with anything having to do with France. "If you don't I will cut off your right arm. Then leg." His English was surprisingly good, with a better accent but less of a vocabulary than Kazem.

In the early days I couldn't predict the course an interrogation session would take. The only thing that was clear was that their evidence was based on emails and their interpretation of ones that they deemed particularly damning. And this didn't follow any set course, although patterns emerged. Hackers contracted by the authorities had gained access to my Gmail account through an unsophisticated phishing scam targeting me and several people I knew.

Through all of my attempts to explain that the Iranian Avocado Quest was *not* some shadowy CIA mission dubbed "Project Avocado," one of the big sticking points for Kazem was the very concept of Kickstarter. He simply refused to acknowledge its existence. "Why would people give other people they don't know money to try to accomplish meaningless things?" he kept asking in the accusatory and rhetorical tone particular to those underinformed and self-righteous ideologues that Iran continues to produce year after year.

"Alan Eyre sent you to Iran, Jason," Kazem proclaimed. "The Great Judge can prove that."

"Please tell the Great Judge to prove it," I responded, but was growing tired of the childish process. They were just trying to wear me down, as if I were a proud bull who had been lanced enough times that he was beginning to stumble.

It went on like this for weeks. The bizarre leaps of logic they would make followed by my attempts to outwit them with my written answers.

"Jason, this is not good. You are trying to be very clever, but the Great Judge is not satisfied. You are an espy and you must write 'I did 'spionage and I am sorry for that.'"

One of the many traits that most English-speaking Iranians share is putting an "es" sound at the beginning of words that start with the letter "S," so "street" becomes "estreet," and then doing the opposite with words that actually do start with an "es" sound, so "especially" becomes "specially."

Also, and inexplicably, "v" and "w" switch sounds, so "water" is "vater" and "Volkswagen" is "Wolksvagen." I learned that through a lifetime with my dad, who always preferred European cars: Wolksvagens, Wolwos, and BMVs.

"Dear Kazem, I am very sorry for anything I might have done wrong, but I did not do 'spionage."

"J, perhaps you don't know what you did. Perhaps you were tricked. I understand this. It happens. But any rookie judge knows that you are an espy and this is 'spionage. You must accept your crimes. Then we can solve your problem. If not, you must be executed," Kazem explained matter-of-factly. "This is the law."

I was dealing with the most hardheaded and least sophisticated people I had ever encountered and they held the keys—literally and figuratively—to the rest of my life.

Kazem's English had other lovable idiosyncrasies. Once, in giving me a lecture about the roots of the Islamic State, he referred to the terrorist network as Saudi Arabia and Israel's "love child." I almost told him that "bastard" fit the situation better but realized if I did that I would never hear him use the term "love child" again, and I was not about to give up that small joy.

I wanted him on my side. Of course we saw each other as enemies, but in keeping it cordial we both knew there might be something to be gained in our own quests—for me to get out, and for him to get something out of me.

"You have no idea what is happening in the world," he told me in an early August interrogation session. "One of your greatest cities, Ferguson"—pronounced *Fair-goo-sewn*—"is on fire."

He took real pleasure in American pain.

"Why do your police kill so many blacks?"

It was one of the stock questions asked of me again and again, as if I were the problem.

Of course I was. To my captors I was America.

But most of the interrogations were cruder, a rapid-fire succession of items pulled haphazardly from random emails.

"Who is Yu Darvish?"

"He's a Major League Baseball player of Iranian origin."

"You're lying."

"No I'm not."

"Why did you bring baseball to Iran?"

"I didn't. I just wrote a story about baseball being played here."

"Why does it matter? This is not a real sport. No one loves it here."

"The guys who play baseball do. They have a federation that existed since well before I ever came to Iran."

He shut up and moved on. Apparently it wasn't an angle worth pressing.

He handed me a printout of one of my emails that had a passage highlighted in yellow.

"Why did you go 'radio silent'?"

"What?"

"You said it right there: 'radio silent.' J, any rookie policeman knows that this is espy language."

"If I was an espy and this is espy language, why would I say something that notifies anyone reading that I am an espy?"

"This is part of your genius. You are espying in our view. It is very dangerous."

"So maybe I'm such a big espy that *I* don't even know it."

"Yes, perhaps. That's what some of us believe."

"You guys will believe anything."

"Come on, J. Give us some credit. We are the intelligence system of the Islamic Republic."

Actually that wasn't true. I had been taken and was being held by the intelligence wing of Iran's Revolutionary Guard Corps, a self-declared intelligence agency run by the former commander of the Basij militia that sees the actual Ministry of Information and Security (MOIS) as their main obstacle. The guys who had me were counterintelligence in the most literal sense: they were completely lacking in intellect. All lifelong Basij volunteers, they were the regime loyalists who grew up with guns and badges.

"I expected a lot more from you guys."

When I told Kazem that it appeared he was doing little more than adding question marks to tidbits he picked out from my personal correspondences, each one harder than the last to respond to with a straight face, he shot back, "J, the American government closed your email account the moment we arrested you. If they hadn't we would have much stronger evidence of your crimes."

"That didn't happen," I said, knowing full well that if my brother knew of my arrest he would have figured out very quickly how to get my Gmail account suspended. I had given him a complete power of attorney to do literally anything on my behalf in the United States, the highest level of trust one can bestow on another private citizen. We don't agree on everything, but that's how confident I am in his commitment to doing the right thing. He'd started in the hours after our arrest, but I wouldn't know that for weeks.

"We also have your phone records. All of the calls and text messages you've made," Kazem told me.

I had heard reports of texts being used as evidence in court but wasn't sure that was something that Iran could actually do. It turned out that all they had access to was a log of mobile-to-mobile communications.

"What is your relation to the Polish ambassador's wife?" Kazem shouted, waving a stapled set of printouts one late afternoon. It was the third session that day. "Tell me right now of your relationship with the Polish ambassador's wife or we will reveal your affair with her!"

It was the most preposterous accusation yet.

Two weeks into my detention I had already been accused of so many offenses it was hard to keep track.

I was a Bahá'í, because one of my alleged "deputies," a low-level producer at CNN, with whom I'd emailed but never met, once published a story about the persecuted religious minority that has long been a target of the mullahs.

Later I was accused of being a sympathizer with the Mojahedin-e-Khalq Organization, referred to interchangeably as the MKO and MEK, an opposition group that is widely reviled in Iran for taking up arms on the side of Saddam Hussein during Iran's eight-year war with Iraq in the 1980s.

The evidence there, oddly, was a 2011 opinion piece in which I argued that the group should *stay* on the State Department's list of terrorist organizations for assassinations they had carried out in the past. In it I wrote that the people of Iran considered the cultish resistance group so traitorous that many adopted the semiofficial name given to it by the Islamic Republic, which is "the Hypocrites."

My membership to an email LISTSERV became a major problem. The Gulf2000 list is administered by Gary Sick, a Columbia University professor and Jimmy Carter's former national security advisor. Its members are a collection of scholars, researchers, journalists, and diplomats, probably including Iranian ones. To my captors it was a message board for spies, with over a thousand people receiving top secret and publicly transmitted documents every day, which were usually just links to newspaper articles.

Based on a photograph found on my wife's computer of her and some female friends on their way to a party, one of them wearing a red sash around her neck, I was accused of fomenting a feminist revolution, because as Kazem very seriously pointed out, "everybody knows that red is the symbol of international feminism."

They even went as far as to bring a Farsi translation

of the Book of Mormon—the religious scripture, not the musical—that I had supposedly commissioned. "We found it on your computer."

"No you didn't," I replied.

Finally one day in his unintentionally hilarious English Kazem asked me if I was "Joe-ish."

Knowing exactly what he meant, I played dumb anyway.

"What's that?" I asked.

"Joe-ish," Kazem repeated. "A follower of the prophets Abraham and Moses."

I considered making a joke about the many Bar Mitzvahs I attended as an eighth grader but thought better of it. That, too, would almost certainly be added to my long list of crimes against the Islamic Republic.

I was of course, though, an agent of Israel, because I was once asked in an email by a Danish friend, who was doing a master's at Columbia Journalism School, about the public perception in Iran of repeated military threats against Tehran by the Israeli right.

I answered that I thought neither the Iranian public nor the state took Israeli rhetoric as a serious threat of attack, and that the mounting sanctions on Iran's economy amounted to an act of war in the public consciousness.

Obviously that friend was a spy because his family

name is Rasmussen (*Rass-moo-sen* in Kazem's accent) and he must have been the former NATO chief Anders Fogh Rasmussen, and if he wasn't actually him, he must have been related to him. "Do you have any idea how common the name Rasmussen is in Denmark?" I asked my interrogators.

Didn't matter. It ended up being one of the main charges against me, a topic I was questioned about for days on end.

This simple exchange became the basis of a charge of espionage for the enemy state of Israel—an offense punishable by hanging—although, as I pointed out in the interrogation room and later in court, everything I had said was the public position of Iran's supreme leader, and therefore the entire Iranian political establishment and its state media.

I was actually doing them a favor by disseminating the state line, but was accused of organizing and conducting a survey that provided "man on the street," or what they referred to as "public square," information to the enemy. Another crime so grave that Iran doesn't even have laws to prosecute it.

I was reminded time and again that the charges against me were incredibly serious. While that was very true, that didn't mean they could be taken seriously.

With each new accusation I provided the best answer I could, which was either "That is not a crime" or "I have no idea what you're asking me," written on a sheet of lined paper with the judiciary's insignia on it, then signed and stamped with a green-inked imprint of my right index finger, for good measure.

"The Great Judge is not satisfied," Kazem would say. "Write again, and this time no 'I believe' or 'perhaps.' You must accept your crimes."

I tried not to falter and instead just kept bobbing and weaving.

"Jason, you have spent a lot of time in Dubai, we know that," Kazem told me during an interrogation. "We know about everything that happens there."

I wondered what he was angling at, but I couldn't guess.

"Come on, Jason, we know about you," he prodded. "You live a dirty life."

Look who's talking, I thought. "Tell me what you think you know about Dubai and me."

"In Dubai there are hotels. And at those hotels there are swimming pools. We know all about it," he said as if he had me cornered.

"Yes, that's very true. There are hotels in Dubai and most of them having swimming pools."

"But we also know about what happens at those swimming pools. There are men and women swimming together." He was getting excited now. "And there is alcohol."

"Very true," I confirmed.

"And what else, Jason? You know. I know you know, I have been there with you."

I was starting to think maybe he had been.

"The beds next to the pools. Why are there beds next to the pools?"

"Umm, so people can get a suntan?"

We both laughed for a second.

"For sex, Jason!" Kazem proclaimed. "They drink alcohol, swim, and have sex. We know all about it."

"I've never been to that hotel, but if I ever get back to Dubai can you give me the name of it?"

If they kill my sarcasm, the terrorists will have won.

"But you have been there, Jason. It is where you met your wife."

"Why are you talking about my wife?" I knew from experience that discussing another man's wife is a red line that supposedly pious people have no right to cross.

"She is your biggest problem," Kazem said.

"Fuck you."

Kazem then explained the latest accusation, the one that got under my skin more than all the rest. It was the idiotic claim that Yegi's and my marriage was a product of a CIA and MI6 arrangement that neither of us was aware of.

"Which one of you geniuses came up with that?"

"It is the Great Judge who believes your marriage is a plot," Kazem explained.

"That is ridiculous. And offensive. I don't know much about Islam, but I know you have no right to question my marriage. It's a sin."

"Yes, you are right, it is. But this is a matter of our national security," Kazem rationalized.

"So that is more important than the word of God?"

"Our national interests are the word of God."

What do you say to that?

"Write about meeting your wife. She will do the same. But if your stories do not match your situation will be much worse."

"Fine."

Again I got to writing. This time a more abridged version of this very real and sweet life event.

4

How I Met Yegi

When I moved to Iran, Barack Obama had been president for just a few months, the country was still reeling from the financial crisis, and moving to Iran to follow my dreams seemed to be the most logical thing I could do. With fifteen thousand dollars in cash that I had stashed away during the last chapter of my life, and doing my best to ignore some of the debt I'd racked up in the months leading up to it, I set out for Tehran, leaving behind my parents, my only brother, and his family, including my two nephews, Walker, who was four at the time, and Paxton, one year old. I knew I'd miss them all, but through experience I had learned an important lesson about family by then: if you have goals in life and want to avoid regret, time with loved ones is best measured in quality, not quantity.

Three weeks after closing shop in San Francisco I was in Tehran, covering the run-up to the 2009 Iranian presidential election, and I was ecstatic. I had several strings—freelance gigs—and I was watching what I could sense was a big story unfolding in front of me. Really, my first one.

During the day I would go to campaign events and press conferences, and at night I would join the carnival on Valiasr, Tehran's main thoroughfare that cuts all the way through the capital for more than ten miles. They say it's the Middle East's longest street.

It divided the city into directions and classes. It was perpetually a mess of congestion and activity. And during those nights leading up to that fateful election, it was the scene of a two-week-long nightly street party, where supporters of the four candidates, and supporters of no candidate—voyeurs and frotteurs—gathered to mingle publicly.

By all estimations no one could remember seeing anything like it. Change was in the air and there was genuine excitement. After four years of the increasingly unpopular populist Mahmoud Ahmadinejad, a former mayor of Tehran who won in 2005 most probably because of low voter turnout—the breeding ground of electoral fraud and nationalism—and a collective rejection of Islamic Republic politics as usual, many Iranians had decided to give electoral participation one more shot.

The atmosphere was charged and the sides extremely polarized. Few other Western journalists had arrived yet, because foreign presidential campaigns aren't news; elections are. The ones who had shown up didn't have press credentials and were busy waiting at the ministry for permission to work.

But I was out on the street filing every day. In a story I wrote the day before the election I argued that the congenial atmosphere of the previous nights wasn't sustainable much longer. The live and uncensored one-on-one debates, unfathomable in previous elections and unrepeatable now, had inadvertently opened up a space for public criticism of the system that hadn't existed before. It was as if Iran was momentarily acting like an open society.

That story got picked up by the *International Herald Tribune* and subsequently published by the *New York Times*. I didn't realize it yet, but that was the beginning of one of the longest Iran reporting runs. From that day forward I was never without a gig. Well, not until July 22, 2014.

It was a rocky road filled with obstacles and challenges, both in the political climate and in my personal life. But somehow I stuck with it, trying, not so methodically, to round out the edges and fill in the color in the general understanding of Iran.

During that time I wrote news; features; cultural stories about art, tourism, food; and opinion pieces. I argued against sanctions. I wrote about everything. There were themes I knew were too granular for any eight-hundred-word news story, so I would fit them in where I could. Unpack the hundreds of articles I wrote from Iran between 2009 and 2014 and you'll find plenty of useful information that previously went unmentioned in international news.

But there also were hiccups.

Several days after the election I received a phone call from a blocked number. In Iran that means the state security machine is calling you. I was told to come to an unmarked building in the bustling center of the city. When I arrived and knocked on the door the guard on the other side asked who I was there to see. When I told him I didn't know, the door opened. Apparently I was in the right place.

In a ninety-minute interrogation with two unnamed intelligence officers I was told that, like other foreign reporters, I was no longer allowed to work and that I should leave the country.

I considered my options. I was having my first taste of journalistic success. I had just rented an apartment. I felt extricated from my old life. I didn't want to leave.

But I was nervous. I knew the story of Roxana Saberi, an Iranian-American journalist who had been imprisoned for several months and released earlier that year, and Maziar Bahari, a *Newsweek* reporter who had just been arrested for covering the street protests.

The story mattered to me, but not as much as my life and liberty did. So I left and went to the closest place possible: Dubai.

I was crashing on the couch of my friend Lara Setrakian, a young and accomplished American television reporter who also covered Iran. We'd met two years earlier and had become fast friends. She had made it clear her door was always open and I took her up on it.

One afternoon Lara asked me to be ready to accompany her to a hotel on Dubai's Sheik Zayed Road. She was going to interview a guy who had been billed as an Iranian opposition figure and who Lara wasn't sure would be able to speak English. She might need some translation help.

We went up to his hotel suite. He was about my age and, as quickly became clear, liked to talk. His

English was good enough for him to explain that his role in the resistance, as he saw it, was to go on Farsi-language channels and encourage his fellow Iranians—though he hadn't visited Iran in over a decade—to join the street protests against their oppressive state. I had seen plenty of people like him before and was unmoved.

As Lara and I began to wrap up our conversation, realizing there was little reason for us stay, the guy made a last-ditch effort to keep us in the room.

"I haven't been to Iran for years, but two of my cousins are in town and they have been active in the protests. They will be back shortly if you would like to talk to them."

Lara and I had already sunk plenty of time into this goose chase so we figured, why not? He poured us each another Chivas and we waited.

It wasn't long before the door opened and a bunch of branded shopping bags entered the suite accompanied by three young Iranian women: two with fair complexions and lacking any kind of discernible energy, and one whose skin seemed to glow from being in the desert sun's rays. That vitamin D exposure seemed to nourish her spirit, too. From the moment she entered the room she took over everything for me.

Her name was Yeganeh, meaning "unique."

Yeganeh sat on the edge of the bed and jumped right into the conversation. She was the only one of the three who spoke English and obviously enjoyed the opportunity to practice with natives. I learned later that we were the first Americans she'd ever met.

Lara asked her questions about the street protests back home in Tehran. Yeganeh got angry, sad, and excited when she talked. She was totally alive and I was mesmerized. I had been instantly attracted to many women before, but the idea of love at first sight seemed ridiculous to me until that moment. I've never doubted it since.

At some point I beckoned Yeganeh from her perch on the corner of the massive bed to come sit next to me on the couch.

Too often in my past I had been shy about revealing an attraction, and I was deathly afraid that this would be one of those moments. She had told us all about her reasons for protesting, about her mother's anxiety and her father's encouragement, his feeling that her generation could correct the mistakes of his and the haphazard handling of the revolution. We talked and talked. Nothing else existed during that time.

I asked her for her email address and I gave her mine. I made her promise to contact me—for strictly professional reasons, of course—the next day, her last before going home to Tehran.

Lara and I got up to leave the room, but I didn't want the moment to end. I lingered, and Yegi and I shared an awkward laugh and even more awkward extended handshake.

When we finally exited to the hotel hallway, the door closing behind us, Lara looked at me. "Whoa," she said, "you're going to marry that girl."

"Yeah," I said. "I think I might."

It wasn't until dusk on the following day that I got a call on my old and very basic Nokia 3110. It was the cousin. I picked it up assuming he wanted me to try to include him in one of Lara's reports. Instead, he invited me to a club with Yegi and the other girls.

I didn't even have to think.

"Sure. What time and where?"

"Come back to my hotel in two hours."

When I arrived, he was sitting with a glass of whiskey in his hand, legs dangled in the pool.

Yeganeh was in the water. I sat down on the end of one of the lounge chairs. Dubai in July is one of the hottest places on earth. When she finally got out of the pool it was dark. I handed her a towel and she patted herself dry. I stared and smiled, not even trying to hide my attention to her every detail.

"I need to get ready. I'll be right back," she said, leaving me with her cousin by the pool.

An hour and a half later she and her sister came back. "Ready?" she asked. "Let's go."

We piled into the backseat of a car that belonged to the cousin's friend, who was driving. He rode shotgun and the two sisters and I sat in the back, Yeganeh sandwiched between her sister and me.

I was feeling bold, and after a couple of minutes in the car I put my hand on Yeganeh's and she grabbed it playfully, and never let go.

In the throbbing club, we found the only spot to sit in the whole place, a narrow and high-backed velour love seat. It was loud, but we could hear each other if we got close.

When the night ended any doubt about whether I would return to Iran vanished. I may not have had a job, but I had a reason. I was in love.

While the street protests raged on in Tehran I weighed the risks of returning. Like most of the correspondents who had been working there I had been told to leave. The rounding up and imprisoning of journalists, dual nationals, and even embassy workers had been enough to keep me away initially, but now the long twice-daily phone calls to Yegi back in Tehran were boosting my confidence. Or at least obscuring my concerns.

I didn't do anything wrong, so I have nothing to fear going back, I told myself, never confident that that even mattered.

I called the Ministry of Culture and Islamic Guidance, "Ershad" for short.

The male receptionist answered the phone. "Salaam, it's Jason calling, may I speak with the ladies?" There were three women who for years were my main point of contact with Iranian officialdom.

I was put on hold and an ice-cream-truck version of Scott Joplin's "The Entertainer"—the theme song from the movie *The Sting*—accompanied my wait, as it had every time I called that office since I first applied for a press pass in 2003.

"Hello, Mr. Jason, how are you today?" a cheery voice said. "We miss you."

"Well, I am okay, but I want to come back to Tehran and would like to know if it is safe for me to do that," I said.

"Of course you can come. This is your home," she said with typical Iranian hospitality. "But you cannot work. And it is not clear when we will give you permission again. So it is better if you don't come."

"I understand," I told her. "So I will be coming back then."

In the fall of 2009 I returned to Tehran after four months away. I didn't have a plan, but there was a woman I had to get to know. Even though we had only spent a few hours in each other's company, when I arrived, we were already together.

My relationship with Yegi very quickly became as serious as was possible in Iran's difficult-to-navigate dating environment. Our time together was mostly spent in my apartment, which was, by all accounts, a dump. But we were happy.

I helped her get a job as a translator at Press TV, which is the Islamic Republic's English-language propaganda network. We hated what they stood for but knew that it was the best job for a young local and non-native English speaker in Tehran: international media was poaching from them and the salary was higher than anywhere else in town. Compared to the companies that hired young educated women like her as interns, using them for months as intellectual slave labor with no intention of compensating them, as all the while male management—usually contacts of the intern's father—sexually harassed Iran's best and brightest, Press TV was a huge step up.

I opened doors for her, but she was always in the driver's seat. I mean I helped her find opportunities and sought her input on everything I did and I never once got behind the wheel of a car in Iran. Our relationship was an anomaly for anyone who took a moment to think about it, but it worked because we both got what we wanted. For Yegi it was permission to have some control and make decisions, and for me it was a strong and dedicated woman in my life. She would say I taught her a lot, but I was the one that was learning the most.

I knew our life together had been tested already, and that our bond was strong, but we hadn't been through anything like this. I wondered if her prison routine was like mine.

The door opened rarely, but often enough that there was no ceremony that went along with it when it did. There were feedings, and a twenty-minute walk in the yard—straight lines, wall to wall, blindfolded, with your head down.

The most thrilling exits were when it was time to see the interrogator. It didn't take long for that warm feeling of anticipation about impending human contact to develop, and when it did it was accompanied by a sinking acceptance that my life had been reduced to this.

It feels as though the interrogations are leading nowhere. If they have anything that looks like legitimate evidence of wrongdoing they haven't produced it. They're still telling me to "just admit it." But admit what? I'm getting tired and detached. That's the scary part. I can feel my grasp on reality loosening. The walls move sometimes in my cell. I know they're not really moving, but it looks like they are. And I don't have any news about my wife.

I never let my mind wander to what they might be doing to her. Well-documented cases of systematic rape by prison staff are part of the Islamic Republic's ugly legacy.

It was three weeks in and I had been completely cut off from the world. My entire reality was the jarring ping-ponging between solitary cell and interrogation room. In the moments that I was being led from one to the other, out of doors, the prison guards spoke more freely; they talked to me in amiable tones.

"Why don't you just cooperate? How long have you been here? I've seen guys much worse than you get out in forty-eight hours," one confided, and then advised, "Just tell them what they want to hear."

Sometimes even Kazem let down his guard, allowing me to enter into his own struggle with the seductive power of the American idea.

He asked me, "How much could someone with a job like mine earn in a good part of America?"

"It depends. I'm not sure how much your American colleagues make, and what do you mean by 'a good part of America'?" I asked, just trying to keep a conversation that wasn't directly about my supposed guilt alive.

"You know, just a normal policeman like me in a place where it is legal to own guns. A good place. Like Texas."

I debated whether to make him feel hopeful or take a crap on his day, knowing that in this part of the world civil servants and law enforcement officers earned very little official pay.

"Oh, you would need about four thousand dollars per month," I said, making up a number knowing full well it would sound astronomical to him.

"This is a lot," he lamented.

"No, it's nothing. Everyone in America earns this much. You didn't think the right to have a gun comes free, did you? We must pay for that," I said, talking down to him temporarily from an imaginary reservoir of exceptionalist American patriotism.

"Yes, I understand." *He clearly doesn't.* "If you are allowed to leave someday, can you get me a visa to go there?"

"It depends on how long you keep me as your guest," I told him.

The informal conversations with the prison staff were exactly the same as the ones I would have on the outside with people I encountered all the time. I learned quickly that, despite hearing years of protestations to the contrary from locals who believed their ruling establishment was run by people who couldn't possibly have shared the same values as them, my captors were just as Iranian as everyone else in this town.

Sometimes the janitors, the guys who saw everything that happened in the prison, were complicit but had nothing to do with it, would call out to me.

"Hey, it's Mister J," one yelled from a distance. "Ask him how his *dodool* is." For some reason, he wanted to know how my penis was doing, using the term schoolkids would.

"Ask him yourself," the guard shot back. "He speaks Farsi."

Not missing a beat, I responded in their direction from under the blindfold, "He's not bad, but he and I both miss our wife."

Everyone in earshot laughed.

Just maintain some semblance of humanity, I kept telling myself.

And then almost within seconds I'd be either back in my tiny cell until who knew when, or in an interrogation room about to be confronted with another printout of an innocuous email, highlighted in orange, which, according to Kazem, had already been translated and sent to the nameless, faceless Great Judge and his supposedly growing file against me.

Islamic justice my *dodool*.

In my cell, I'd run out of things to think about. God forbid you had a curious mind that asked questions, as I do. Not only was there no right to ask them—a commandment that had been drilled into my head starting with those maddening first days of state-sponsored disorientation—but everything I didn't even realize was available to me at all hours of the day when I was free, with the exception of breath, had been robbed from me.

There's no Google search in prison. I was stripped of my right to information. It quickly became the highest form of deprivation.

I know you want to read about what it's like to be isolated in one of the world's most infamous prisons, how someone survives that, but you don't really want to know. It's a hard experience, designed to dehumanize and disjoint the subject from reality, and guess what? It works.

If you haven't spent significant time in prison you've missed one of the essential aspects of the human social experience. You're lesser because of it. At least that's what I kept telling an imaginary audience while I struggled through each day.

The world becomes a hybrid of the life in your memories turned fantasies, the one you used to inhabit, and this, the life you never imagined you'd be forced to endure.

Another thing you have in prison is nothing to do, so you think. Think about everything. You pull up conversations from every era of your life, sometimes with people who are long dead, and you conduct a re-examination of key decisions. Done in that sort of tight vacuum, such exercises are emotionally precarious.

Old friends, people you haven't thought of in years, dumb shit you wish you never said and desperately want to take back.

I made a lot of mental lists. Places I've been and ones I'd like to visit. Favorite restaurants. The home-run kings in each league during the years of my childhood.

Sex came up often, probably because being detained in isolation brings one as close to mortality as anything that isn't attacking your physical body can do. You replay a lifetime of sexual encounters—in retrospect there is no bad sex.

You think about every sex act you can remember and some you can't. Sex you should have had and didn't. Sex you had but realize you shouldn't have. Actually, scratch that. In prison you realize that all sex between consenting adults is worth it. Sex that was real and some that was imagined, that came to you easily or that you had to earn.

It was sort of hard to avoid as a topic in the particular prison I was in, as so much of the interrogators' world-views were skewed by all the sex they weren't having and deeply wanted to believe you were.

5

Writing My Memoir

The interrogation sessions were leading nowhere, it seemed. No one was satisfied. Not the Great Judge, not Kazem, and least of all, me. All they were trying to get me to do was the one thing that I refused to do: admit that I was knowingly or unknowingly spying for America.

Sure, I could have said something like "Because the *Washington Post* is the paper of record for the capital city of the U.S. and read by many people, including possibly government officials, I was unknowingly providing legally gathered information from Iran to them through my public news stories." But I wasn't about to say "I was sent by the U.S. government to report sensitive and classified information from Iran under the guise of being a journalist."

I guess they thought they would break me down to that point, and yes, a few times I got close, but the longer it drew on and the weaker I got, it just seemed too farcical. There was a total lack of consistency in everything that I was told. Yes, such a process is designed to confuse, but not to be utter nonsense. How could that be productive?

Well, it wasn't.

During one session Kazem actually told me that my messy Gmail inbox was "a sign of my agility as an espy," that I had been trained to keep it so disorganized. It was around that time that, even in the face of threats of execution by beheading—the supposed Koranic punishment for traitors—I started rolling my eyes.

I was afraid, no doubt, but more that they would run out of other options. So I did the best I could to cooperate, providing as much information as I could, and usually this took the form of explaining a worldview and life experience that was absolutely alien to them. It was exactly what they wanted, and ultimately, I suppose, what they needed.

But I couldn't explain that to them. They were too dumb and one-dimensional to get it.

One evening the door to my cell opened. It was Kazem. It was the first time he had come inside the cell block to see me. I was worried, because inside prison one realizes quickly that surprises are rarely a good thing.

"Here is paper and pen," Kazem said, handing me familiar tools. "Write everything about your life," he told me.

"Like what?" I was not in a mood for guessing games.

"Everything that you think is important," he answered. "From the beginning."

"Why?"

"Many people think you should be executed. Including me. You must give us a reason not to. Although you have done very bad things, every man must have an opportunity to be saved. I should," he said calmly, "no, I must try to save you."

I had no idea what he was talking about, but armed with tools I knew how to use and a lot of time, I was no longer alone. I stretched out on the floor of my cell, folding my two blankets and placing them under my torso. It was the closest approximation to a desk in that environment that I could muster. I lay down and started to write.

From the very beginning my life has been another episode in a long series of improbabilities. That's not really that unique. I'm American, the product of parents who were raised in very different but equally conservative settings and got out.

My brother was born in 1970, and then so was I, on the Ides of March of America's bicentennial year, 1976. The Year of the Dragon on the Chinese Zodiac. By all accounts I was a happy kid in a happy family.

My dad's business, Rezaian Persian Rugs, thrived. A year after I was born he remodeled his shop front to resemble a Persian dome, becoming an instant landmark at the beginning of a well-traveled strip of Highway 1 that people all over the world talk about even now. It still stands, and it's still a rug shop, but it's been a long time since it was Rezaian Persian Rugs.

We lived in a large two-acre compound in San Rafael, fifteen miles north of San Francisco. My parents bought it in 1972, before it was a desirable location and around the time the town served as a shooting location for *American Graffiti.*

Everything about our home was big. A five-bedroom house with an eighteen-hundred-square-foot guesthouse, both covered in Persian rugs. We had the biggest home pool I have ever seen.

My parents loved the openness and seclusion of the land with space for horses and fruit trees, and the privacy we had as the property butted up against county land that still hasn't been developed.

The door was always open and weekends meant large gatherings of relatives and friends by the pool in "Taghi's garden," barbecuing and laughing.

We were less than a mile from the Marin County fairgrounds and on the Fourth of July everyone would climb to the roof to watch the fireworks display that danced above the patch of massive eucalyptus trees on the eastern border of our spread.

For my dad his entire identity became wrapped up in that piece of land, and for us it was an idyllic place to grow up.

My brother, Ali, and I went to private school, where we fit in just fine. I can't recall a single instance of anything even resembling discriminatory treatment from classmates or teachers. We were different from many of the other kids only in that we weren't Jewish and had just one set of parents.

I made friends easily, which is good, because I was always the fat kid.

I loved Star Wars. That was my thing. All the action figures. Star Wars birthday cakes, Halloween costumes, bedsheets. Even wallpaper in Ali's and my bedroom. No one else had all those.

Later I fell in love with baseball, specifically the Oakland A's. Aunt Mae and Unk, who had helped raise my mom, moved into the guesthouse—"the cottage," we called it—when I was five. Unk, who was in his midseventies by then, became my closest friend, and he and I had season tickets starting in the mideighties, just before the A's got good. We went to an All-Star game, the playoffs, and two World Series.

We were typical Americans with a few Iranian twists.

Three of my dad's six sisters and his younger brother had all moved to Marin by the early eighties. My grandparents, who had come from Iran for an extended visit before the shah had been toppled, stayed. More of their children were now in Marin County than were in Mashhad. Twice a week we would gather at my aunts' home for dinner, learning the initial lessons of the culture through our stomachs.

On the Persian New Year, or Nowruz, the first day of spring, we would all get together just like we would on Christmas or Thanksgiving. Our life was a quintessentially blended one and that suited me fine.

I remember quite vividly the visceral responses of my mild-mannered and fun-loving aunts and cousins who would curse the television every time reports of fuming Iranians burning American flags were aired.

"That is not who we are," they would plead, as if someone inside the set were listening, waiting to be convinced.

And from my point of view they were right. "Iranian" to me meant vibrant, talkative, generous, colorful people who liked to dance, wore too much makeup and perfume, and had funny accents.

The early 1980s had not been good to my dad. Shortly after the Iranian revolution, when any endeavor remotely involved with Iran or Iranians was considered suspect or taboo, he helped usher in a new and enduring stereotype: the never-ending rug shop liquidation sale.

It was 1984 when my dad started his first one, and it worked.

The haze of the cocaine-and-marijuana-infused seventies was wearing off, and even in ultraliberal end of the road Marin County, the bigotry of America against a new boogeyman worked to devastating effect. Following the euphoria of the fifty-two U.S. embassy hostages' release in early 1981 for my dad and other Iranians in America, life would never be as it was.

"The Sale," as it forever became known in local rug man lore, was his chance to settle the score. For well over a year the signs hung in the window and the radio ads pounded the "all-news, all the time" and classical-music airwaves. The advertising men said that was his target audience, because none of the online tools to quantify an ad campaign's efficacy existed yet.

A phenomenon that has devolved into a stereotype of popular culture started off as an event.

Customers drove from hundreds of miles away to buy rugs at "up to ninety percent off!," because in those days people still listened to the far-reaching waves of AM radio. There were deals to be had. Friends who had never considered buying a rug walked home with ten. They thought he was doing them a favor. Little did they know it was the other way around.

Searchlights lit up the sky. I had only ever seen those outside of a circus before, but these were drawing the people to us. I say "us" because on some weekends my brother and I became part of the sales force. The fish were biting and Dad needed as many lines as he could get in the water. I remember the buzz of shaking hands to close a deal, and the exchange of stacks of hundred-dollar bills. Selling rugs, like selling drugs, is often a cash game.

It's impossible to know how much Rezaian Persian Rugs did in sales during the liquidation, because due to industry norms the books were notoriously incomplete, but it was in the multimillions. For a nine-year-old kid who was suddenly recognizable locally, it was thrilling. And then it stopped.

I didn't understand what bankruptcy was or how someone got to be called a creditor, I just knew my dad had a lot of them, and it was messy.

For several years after that Dad tried his hand at different businesses. He opened a shoe store in San Francisco—Shoes, Shoes and More—and later the first Persian restaurant, Shiraz, in Marin County. But rugs were in his blood.

On weekends before I could drive I would often end up at flea markets and estate sales with him. He enticed me with the possibility of baseball cards, which I collected passionately from the time I was seven. We'd pay the entrance fee, he wandering in search of rugs and I looking for cards. Once we found what we were looking for his opening offer was a standard "How much for all of them?"

We ended up with a lot of baseball cards and a lot of rugs. Decades later I'm still trying to figure out ways to unload them all.

As I grew up he picked up an appreciation for baseball, accompanying me to games when Unk got too old—he'd wear an all-denim ensemble with a mesh Oakland A's cap, and was invariably confused for a migrant worker on a night off. As he was almost everywhere, at the ballpark, Taghi Rezaian was an anomaly.

He'd have a black coffee and a hot dog, and fill a cup with sauerkraut, which he'd eat as though it were

a side salad. He grew to understand and love baseball—
further evidence of his Americanness.

In the years after his bankruptcy, though, and until
he could start building credit again, he was very restless and these excursions did little to distract him.

His experience was gold to a lot of rug men. In those
days it was easy enough to get a permit in most communities to run a liquidation sale. He knew where you
could and could not do one, depending on county regu-
lations, for a very wide radius. He had a nose for where
a sale would work and where one wouldn't. It was just
one more of the many formulas—equal parts median
income, common sense, and gut feeling—that he had
unscientifically worked out to rationalize a project.
Some worked, others didn't.

Election years, he would tell you, "are very good
for sales." Unless they aren't. When gas prices are high
sales are low, because "people won't drive as far as they
used to for a deal." Retail businesses must stay open on
Super Bowl Sunday because "it's a great day for women
to shop."

There was always an answer and it didn't always line
up with reality. Opacity was an industry-wide problem
that chipped away at the rug community's credibility
until it became the joke that it is today.

That stereotype stuck, and I always considered it a great shame, not because it wasn't deserved—it most certainly was—but because the Persian rug man, although never overly respected, was always the best link to the old country. For one thing, they continued to do business with Iran even when it was illegal, figuring out new and inventive ways to skirt U.S. embargoes on rugs. That alone kept them connected to the beat of commercial life in Tehran from the other side of the world. And in a country where commerce is everything, those who understood what was happening in the bazaar provided the best analyses of events on the ground.

True, they weren't married to facts and figures, but statistics coming from Iran and other Middle Eastern countries are notoriously flawed. The rug man's information, like that of a good reporter, was gathered at the street level from a variety of sources: exporters, housewives, transport professionals, weavers, repairmen, trade ministers. A cross section of the society.

Maybe that's why from the mid-1970s until he died, my dad was always on the list of people to consult about Iran, whether it was local media, members of Congress, or federal agents, who started appearing at our home in Marin periodically, unannounced, in 2007. They were trying to retrieve a former agent named Bob Levinson who had disappeared on an Iranian island. At the time of writing, he's still missing. For some reason the Feds thought my dad could help.

It wasn't that he was a "person of interest," but rather he was someone who could provide color, a granular read on Iran, beyond headlines and bombastic statements made by Mahmoud Ahmadinejad and others.

As the years passed and I spent more time in Iran, but also in my dad's rug shop, that legacy was being passed to me. All of it.

Revisiting where I came from felt good. *Someday this will also be just a story. One that someone else gets to tell.* The tiny act of being able to write my own story as my captors tried to rewrite the realities of my life was a weapon. It offered me an escape from my present circumstances and the sliver of perspective. It was a gift that they hadn't meant to give me but couldn't take back.

Writing about my family history, in that great void where I couldn't touch or hear anyone I loved, was soothing. By then comfort was all I was looking for.

"The Great Judge says it is not good," Kazem told me solemnly two days after I had given him the pages.

Tough editor, that Great Judge, I thought.

"It is not complete."

"How do you know? It's *my* life story."

"There is nothing about your 'spionage."

"Because there is no 'spionage."

"The Great Judge wants to know more about where you have traveled. He says tell him about Thailand. Specially any negotiations with Israeli agents there."

"Bring me more paper."

At my core I'm a travel writer. I wrote about everywhere I'd been and why I'd gone. Maybe that would put an end to all of this.

Just over a month into my time, Kazem came back and told me that we were going for a trip.

"Where?"

"Court."

"Where is my lawyer?"

"You don't have the right to a lawyer yet," he said as calmly as always.

I sat in a waiting room, where I was brought a glass of tea, and then was taken into the office of Tehran's prosecutor, Abbas Jafari Dolatabadi, considered by most Iranians a caricature of the most sinister members of the Islamic Republic's legal establishment.

He was the farcical character you might expect. He dressed in a shabby gray suit over a black shirt— "perpetual mourner," which is a look in Iran—with a large ashy bump on his forehead, or the "stamp" of prayer, one of the most clichéd revolutionary bona fides, and his missing hand, presumably lost on the battlefield in the war with Iraq and undoubtedly his highest qualification for holding this influential post as long as he had. Another possible badge of honor was his 2011 designation by the United States as a human rights abuser.

I had seen him once before in 2013 at the Interior Ministry in Tehran when he, along with hundreds of other Iranians, flirted with a presidential run. I wrote a story about the carnival scene there, which I headlined "Want to Be the Next President of Iran? Take a Number."

He wasted no time in launching into questions about me and my work. It seemed as though I was finally making some progress, as he was the first person I'd seen since my arrest that wasn't addressing me as a spy. But that's probably because he wasn't really talking except for his very short questions.

"Why did you make the 'Happy' video?" he asked.

"What?" This was the first time I was hearing this particular accusation.

"You directed this 'Happy' video. That is a big problem for you," the prosecutor told me. "It's why you are here."

A few weeks before my arrest there had been a brief hullabaloo about an Iranian-produced video clip for the Pharrell Williams song "Happy." Fans all over the world shot amateur video of people in their local countries dancing to the hit tune. It was a viral thing on the Internet and some industrious young Iranians had gotten in on the act—and were quickly arrested for antiregime activities. I didn't see this as a big story, or at least not the kind that was worth a lot of space, but my editors did, citing an interest from our audience. So I wrote it up, significantly later than other outlets. Now, a couple of months later, the clip's production was being pinned on me.

I resisted.

"That had nothing to do with me. If that's what this is all about it's time to let me go."

Dolatabadi wasn't the type to get very animated. In fact he seemed disinterested. If I shut down a line of questioning he wasn't persistent. He just moved onto the next thing, and there was a ton of them.

Finally he, too, got into the avocado act.

Jesus Christ, I was thinking. The fact that my failed avocado project ever got legs as an accusation is so far beyond far-fetched that I struggle with whether I should even bring it up in retrospect.

After I attempted to explain Kickstarter and crowd-funding one more time, he cut me off.

"At least tell us where the avocado farm is."

He can't be serious. I shook my head and rolled my eyes. *Hard to tell with this guy.*

He was irritated. This is a typical look on Iranian bureaucrats. I knew it well. It was August and the room was hot and smelled of one too many goons who hadn't showered in days.

"I'm not convinced," he concluded after an hour and a half of abrupt backs-and-forths.

"Of what?" I asked.

"That you are reliable. I don't have enough *trust.*" He said the last word in English. "Maybe next time."

"When will that be?"

"Only God knows."

And with that I was blindfolded again, put back in the van, and redeposited in solitary.

After weeks of getting nowhere with the truth I tried to switch gears. The only goal I had at that point was to get out of solitary confinement. I understood, to some degree, the case they were trying to build against me. There was no backing up what they accused me of, but in the vacuum there was no way to disprove it either.

I labored long and hard. *What's the right thing to*

do here? They see me as representative of the enemy strictly because of my employer.

"I'm an officer in the West's soft war against Iran," I wrote. I felt disgusted with myself, but also relief.

Kazem was excited.

"This is very good." It was exactly what he wanted.

But he came back the next day dejected and a little angry.

"The Great Judge says you can't say that," Kazem said.

"Why not?"

"It sounds too much like our way of talking."

That's what I was going for.

But it worked. Sort of.

The next day Kazem told me at the end of a long session that I was being rewarded for my cooperation. It was late August and the days were getting shorter.

"You do something for us, we do something for you. This is our way," he said arrogantly. "Today you will see your wife."

"You're lying," I said, but I was desperate to have some reason to hope. "When?"

"Now," he said. I didn't believe him. "Prepare yourself. Be a man. Do not cry."

"This is really happening?"

"Yes. Why not?" As if I'd only imagined the past month.

He led me into an interrogation room that I knew well.

"Sit here and do not talk. You can only see each other. No speaking."

Even in these circumstances that seemed an excessive demand, but it wasn't a moment for me to try to make deals.

Yegi's interrogator Siamak arrived and said, "She will be right here," as if he were my friend. I was anxious and they were oddly giddy.

A middle-aged woman in black chador walked into the room ahead of Yegi, who was wearing a pink version of the same kind of prison pajamas I was wearing and a veil. She was blindfolded. They sat her across the room, facing me.

"You can uncover your eyes, Miss Salehi," Kazem said.

I prepared myself with the biggest smile I could muster. Her eyes adjusted and she burst into tears at the sight of me. She was at once happier and more scared than I had ever seen her. Maybe not scared, but horrified. Her jolly, heavyset husband was now a shadow of his former self.

She smiled and tried to keep her composure and then motioned for me to stand up and turn sideways. She wanted to see the profile view. I was, it turned out, as thin as I thought myself to be. Maybe thinner.

They allowed us four minutes together and then they led her away.

As she left she turned and said, "Jason, I want to have a baby."

I smiled and told her, "So do I."

I went from the greatest moment of my life, one that I wanted to just savor, to being alone again in the room with Kazem.

"How was it?"

I didn't say anything and began to weep the tears I wasn't allowed to release in my wife's view.

"You did well," he said. "I commend you."

And then he led me back to my cell.

Seeing Yegi for the first time began the period of tiny incentives. After batting me with the stick of threats for a month they began feeding me carrots.

"Jason, you are a Muslim," Kazem proclaimed several weeks into my detention, "but you don't pray. This is very bad."

To the question of my faith or lack thereof, I've always faltered when trying to answer.

My lone visit to Mecca, with many of my Iranian relatives and my dad in 2006, did more to drive me away than anything else could from practicing organized religion. But I realized it was definitely not an argument I was going win in Evin. So I played along.

"How can I pray when I've never even read the Koran and you won't give me one in English?"

"Would you like to read it?" Kazem asked.

"Very much so." It was true. I was dying for something—anything—to read.

"I will try to get permission for an English one," he promised.

"You need permission to let me read the Koran?" I asked indignantly. "That is not justice."

"You are right." I think he actually meant it.

Before praying, Kazem told me, I had to learn the way to ritually wash myself. This had to be done every time one defecated, urinated, or had sex.

"Or when you sleep and it's like you had sex," he explained. It took me a minute to realize he was talking about wet dreams, the only sexual experience he was conceding that he'd endured. He had a real contempt for all things carnal, which was probably just masking a fascination for the same.

He showed me, without water, how one performs the

ritual ablution, or *ghosl*. I learned it quickly. It wasn't very complicated.

He was obviously proud that he'd gotten me washing myself correctly and on day thirty-five he presented me with a very ornate Koran translated into Farsi and English alongside the original Arabic.

"Please treasure this forever. If you leave here to go to another prison or go home, please take it and remember me."

"No matter how hard I try," I assured him, "I'm sure I won't be able to forget you."

"It was very difficult to get permission, but I told my chief that it is every man's right to read Koran. Like you said. He agreed."

I read it cover to cover. For now let's just say a man's faith should remain between him and his maker, if he believes in one.

You should be very happy. This time as our guest will make you famous," Kazem announced one day when he didn't have many questions. "If you are ever released you will be rich and I will still only be a policeman."

"Is that what you call yourself?" I wondered out loud. "A policeman?"

"It is my job. I have many jobs," he said, pulling

out a wallet filled with many different ID cards; one of them was indeed a police officer's with his photo on it. He wouldn't let me see the name. That made sense, but even if he had, I wouldn't have been able to read it. I was still completely illiterate in Farsi.

"Well, I don't think there will be any movie. There's already been at least one about this place," I remarked.

"Yes, *Rosewater*. It is coming soon," Kazem announced. "Do you know him? Maziar Bahari?"

"No." I was telling the truth. "Never met him."

"We treated him very well here. But now he says we torched him," Kazem lamented.

"Tortured," I corrected him.

"What?"

"You said he said you torched him; that means you put him on fire. You meant torture. Like beating him."

"Yes. Torture. Thank you. It is very good practice for my English to be with you. He said we tortured him, but it's not true. I know his expert. He is my friend. Every day when he came to work he told his wife, 'Make food for two people, not just one.' And now he says bad things about us. It's his fault I cannot bring you food from home."

I think he's fucking with me, I thought. *I am so hungry.*

"Yes, we treated him even better than we are treating you."

"That wouldn't have been too difficult," I told him.

"Be fair, J. For all that you have done against our system you should be dead, but we treat you well. Here is like hotel. A nice quiet room. It's very peaceful." *Yeah, he's definitely fucking with me.* "You are upset today. But be hopeful. You are a very important people."

I just stared at him.

"Come on, who will be you in the movie?" he prodded.

"If they make a movie I know exactly who will play me," I said with extreme confidence.

"Tell me."

"Denzel Washington." It was my semisarcastic answer to a dumb question that everyone has been asked at least once.

"Which one is he?" Kazem asked.

"I'm sure you've seen him many times." I thought of all the films he'd been in and the ones Kazem might have seen. *Glory? American Gangster? The Hurricane?* I was pretty sure he wouldn't have seen any of those.

"*Training Day,*" I said.

"Yes. I know it. The one about the bad black policeman," Kazem said, summarizing the plot. "But you don't look like him. You are too fat. That man is handsome and he has hair." Kazem was talking about Ethan Hawke.

"No no. I mean the black one."

"Oh, Malcolm X?" Kazem was puzzled.

"Exactly." I had forgotten that one. Of course Kazem would know *Malcolm X*. There were billboards of his grandson, Malcolm Shabazz, all over Tehran then, calling him a martyr; he had been murdered in Mexico that May. Malcolm X was lionized in Iran, and so his biopic, starring Denzel Washington, was replayed often.

"But you are not black." Kazem was confused.

"It's my damn movie, I can have anyone I want play me. And I can get you your choice, too. Who do you want to play you, Kazem?" I was just trying to have a little fun.

He thought for a moment.

"I want the bad boy."

"Who's the bad boy?"

"The movie. Two blacks. *Bad Boys.* They are police like me."

"Oh, you want Martin Lawrence to play you?" Now I was fucking with him.

"I don't know his name. The bad boy."

I stuck my hand behind my ears and pushed them out a little and did my best Sheneneh. "Yeah, I guess you do kind of look like Martin."

"I don't want the clown. I want *Independence Day.*"

"Ohhhh, Will Smith. Okay, this could work, but we may have to set the movie in the Caribbean or Nigeria."

"Yes. They don't look so Iranian."

We both laughed. It was a moment that stuck with me. I now knew a little bit about Kazem's taste in movies, and from that I might be able to extrapolate other information. I had to stop thinking of him as simply an Islamist adversary to my American sensibilities. I had to find the points of connection.

And I recognized he was doing the same thing with me.

6
Circumstances Change with the Season

September 4, 2014

Earlier than the usual time for an interrogation, the door of my cell swung open. It was a Tuesday morning. Forty-two nights in solitary notched on the wall. Six weeks. *Six fucking weeks.*

The guard was one of the nicer ones, a guy in his twenties, shorter than me, who recited passages from the Koran and had tried to teach me one about asking for forgiveness that he told me he said whenever he felt helpless. It hadn't worked for me.

"What's happening?" I asked him as he led me down a corridor out a side of the building I didn't recognize.

"You're going to court. I think it's finished," he proclaimed.

"Finished? What do you mean?" I asked, bewildered. "My trial?"

"No. I think you're being released," he whispered excitedly. "If we have done anything to make it difficult for you, please forgive us."

"Of course!" I told him as my natural optimism began to kick in. "You have been very kind." I had no reason to believe him, but I didn't have any reason not to either.

I was back on a dirt road. I thought I recognized a slope from the night we were hauled in. I was flanked on either side by men who didn't talk and held me by my arms. They were forceful but remained calm, which made the situation feel all the more sinister. It was nothing like I'd imagined hostage scenes to be: chaotic and filled with threats against a struggling victim. They had done this before. Often. And what was I going to rebel against?

They guided me into the back of a van, to the seat directly behind the driver's.

A guard sat next to me and another behind me. He put his hand on my shoulder, as if to say, "I'm here."

Underneath a corner of my blindfold I could see a flap of black cloth draping toward the floor, which was almost certainly a chador, the all-encompassing sheet worn by many pious Iranian women.

Behind the woman I could hear someone else breathing and feel the anxiety she brought with her. It was another woman, but I had no way of knowing who. We weren't allowed to talk.

I had to know if it was Yegi. When all doors were closed and the engine was turned on I said, "Salaam," offering greetings to whoever was there, knowing someone had to answer. That's customary.

She let out a tiny cry, and I knew then that I was with my wife.

I kept talking and the driver—the same bastard who drove us to prison that first night—told me I needed to stop.

"She's my wife," I responded.

"You'll have time to talk when we get there," he responded.

I wasn't sure whether I could trust this or not, but I did as I was told.

Yegi cried behind me, and for the first time in the five years we were together she began reciting prayers.

After twenty minutes at high speeds on a highway headed mostly downhill, we began to wind around the hectic midmorning traffic of central Tehran.

Out the tinted window I saw a silly little train on wheels ferrying passengers around carless roads. We were approaching Tehran's bazaar. I had spent so much time there over the years, from my first visit to Iran in 2001 to all the times I guided foreign visitors through the tiny covered alleys of rug shops and the various restaurants in and around the ancient shopping mall. I went there to work on stories and just for fun. It was one of my favorite parts of Tehran. It made Istanbul's Grand Bazaar feel like a Disney version of a Middle Eastern souk. There was no map or television screens announcing the local weather, international news headlines, or currency exchange rates. It was all grit and commerce and I loved it. And as we pulled up that morning I understood that the odds of my ever seeing it again as a free man were slipping away and there was nothing I could do about it.

The van pulled into an alley and then passed a security gate. We were back at Tehran's media court, where we had been processed six weeks earlier.

Yegi got out of the van behind me, and they led us in staggered, so we couldn't hold each other's hand. Court was apparently in session and the hallways were bustling with activity. We were the only people dressed in prison clothes and everyone stared.

It was, until that point in time, the most humiliating moment of my life. And it got worse.

Stumbling weakly down the hallway, I could see Yegi's parents running toward us, and behind her my uncle, my dad's only living brother, his former business partner, whom he had been estranged from for the final years of his life.

A wave of shame overtook me as I anticipated a harsh reaction from all sides of the family. Leading to that moment the sense of disappointment in us that I was sure our families felt had kept me up at night. But by the time we walked out later that afternoon it was one less thing I had to worry about.

Waiting inside the office was my sister-in-law, Taraneh. Almost immediately she announced, "The whole world is with you. Everyone knows you didn't do it."

"Do what?" We still had no idea what was being said about us publicly. In fact, they were still telling me some days that I had been reported dead in a car accident.

Right then the wall of lies started to crumble and there would be no putting it back up.

"John Kerry has spoken about you twice," Taraneh told me.

"I did an interview with the *Washington Post*," Yegi's mom said, shaking her head and putting her hand on her face, as if to say she was embarrassed. "They called me on the phone and I talked to them." I wondered which of my colleagues had made that call.

Taraneh's eyes got big and she said, "*Washington Post*," clasping one hand behind the other to say without saying that my employer had our backs.

It was the best feeling I'd had in weeks.

My mother-in-law tried to feed us—she had brought along nuts and a couple of bananas—but I couldn't eat. That was a sign for these people that something was wrong.

"You're so thin," she said, with fear in her voice. "Are they feeding you?"

I made a joke right then, playing on the fact that in Farsi the word "regime" refers to a diet in addition to having the same meaning as it does everywhere, a negative way of referring to a political order. I noted that anyone who thinks the Islamic Republic "regime" is anything but the best in the world is wrong. With this regime I'd lost forty pounds in forty days. Everyone laughed the dreamlike chuckle that is reserved for encounters with loved ones who are temporarily back from the dead.

We sat in that room for half an hour trying to catch up on life. I asked questions and my in-laws answered the ones they were allowed to; others made our IRGC chaperones threaten to cut the meeting short. A secretary in a chador looked up from her work from time to time, looking worried, as if she too might be implicated in our case simply because she was forced to host this bizarre meeting in her office.

Midvisit Yegi was called into the judge's office. It was Alizadeh, the same young and dapper—and for those reasons out-of-place, really—judge who had signed our arrest warrants and processed us six weeks earlier. I was certain our nightmare was ending.

I continued chatting with the family, trying my best to put on a strong face.

"How bad is it?" my mother in-law asked.

"Oh, it's not bad at all. My interrogator is a very nice guy. We joke all the time. He says if they ever make a movie of my prison time he wants Will Smith to play him." Everyone laughed.

My pragmatic uncle repeated a refrain every time we made eye contact: "Don't sign anything," he would say under his breath. As though this were day one and I hadn't signed a thousand pages of lies and half-truths that were at that moment being twisted into my plotting to take down the Islamic Republic. Their collective ignorance about what we were being put through was quaint, and not in an adorable way.

It was my turn to go in the room. The judge allowed Yegi to stay, presumably to help translate for me. We sat next to each other clutching each other's hands, forbidden to speak in English, denied a lawyer, as we were told, once again, we didn't have that right yet.

On Alizadeh's desk were stacks of documents in the blue and pink folders that are used in every Iranian government office. They were Yegi's and my interrogations, which we had been forced to write down, and in my case as they were originally in English, there were also "official" translations.

As Alizadeh read through the contents of a summary sheet of my supposed crimes, for the first time I began to see, in practice, how it was possible to take innocuous details and twist them into admissions of guilt for crimes so nefarious they don't even exist.

The *Washington Post*'s being the paper of record for the nation's capital and the possibility that it might make its way to the president's desk became a major charge against me.

But there were lesser charges, too. I wrote for websites on political topics, I met with officials of multiple countries, I had attended a round of the nuclear talks. All true and all legal by all standards.

That session ended, and when I asked Alizadeh for my constitutionally protected right to be released on bail I was denied.

"For now the investigation must continue," he replied.

They had absolutely nothing they could convict us of, and so they just held us.

We were back in our cells in time for a late lunch.

It'll sound obvious, but one thing there's plenty of in prison is time. But when the imprisonment is a hostage-taking wrapped in judicial packaging, as it was in my case, time becomes the captors' faceless accomplice as there is no clear picture of when the abuse, manipulation, and shameless lying will end, or if they will. I was told repeatedly that my case was special and resolving it would be difficult. Over time I understood that to mean "We've got no actual case against you, need to come up with something plausible, can't, and have no exit strategy."

This is what I told myself as I waited for the door to open and my next breath of human contact, however antagonistic it was, to save me from drowning.

The preliminary investigation stage of my detention was the biggest personal test I'd ever faced, but I knew there were more difficult fates: tighter quarters, beatings, more active forms of sleep deprivation, starvation. Waterboarding. Was I being tortured? And if I was, why wasn't I more sure of it?

It took weeks for it to set in that solitary is the essence of psychological torture. It's designed to confuse, which, when intended to induce weakness or compliance or to gain information or leverage by force, is the epitome of torture. It's why the use of solitary in interrogations is often grounds for throwing out a case. Under no circumstances is long-term solitary confinement justified. There is nothing just about it.

But as you're living through it, the thought of how unfair your current predicament is is only an abstract concept designed to help you pass away the hours. There is nothing that can speed up that time, and your captors know it. To a certain degree, you will break. Everyone does. The key is to not give away the farm—or at least convince them you aren't—in the process.

The fear and constant sense of heightened anxiety creep up on you quickly. It's not easy, but maintaining a sense of humor becomes essential for survival. Besides making it just a little easier to cope, laughter and the knowledge that you can still muster a little mirth, and elicit even more from those around you, provide perspective.

It's the self-doubt, though. That's the killer.

Cracking jokes—oftentimes to an audience of one: me—became my only way to combat it.

Having a sense of mortality is a good thing, and as time passed I developed a system-override mechanism that told me, "Hey, pal, you're gonna die. Just not in here."

But permanent damage had already been done. I wasn't the same person who had been dragged in seven weeks earlier.

My body was a different one. I had lost so much weight. In all honesty, I was glad to lose it, but I was suffering from new ailments. Headaches, eye infections, pains in my groin. All of it attributable to the prolonged stress of that epoch of deprivation.

Worse than that, though, I didn't fully recognize the guy inside my head. I was perpetually scared of what came next. But also becoming resigned to it.

I remembered a conversation I once had with an Iranian journalist friend who moved to Berkeley after spending his fair share of time in Evin. "If you ever get arrested in Iran there are two things to remember. The food isn't *that* bad and you *will* get out." It had become a mantra.

Then, very suddenly, my life in Evin took its biggest turn, on September 10, 2014. Kazem told me at the end of an interrogation that I would be coming out of solitary into a shared cell. I was apprehensive. *Is he lying? If not, who are they going to stick with me? Will he be a mole? Is it going to be worse than being alone?* I had no answers, and it didn't really matter, because I had no choices either.

At around eight P.M. a guard came to my cell's door and told me it was time to go.

I had my possessions in hand: a couple of blankets, my Koran, and my toothbrush; it felt like I was moving across the Atlantic when he told me to put on my blindfold. Where we would have turned right to go to the prison infirmary, we turned left.

We walked through a short corridor that I could feel was brightly lit, then outdoors again, and stopped a few steps away under a small entryway. "You can uncover your eyes," a new voice told me. He was a guy in his forties, sturdier than any of the other guards I'd seen and better dressed, with very thick glasses. After a minute or so another prisoner arrived with the same guard who brought me. He was nervous and emotional.

The guard spoke to him condescendingly, but the guy obviously couldn't understand Farsi. That much was clear. The new guard, the one in the thick glasses, said a couple of things to him in his language and the prisoner lit up.

"Tell him that everything is going to be all right. I'm better than Seyed," I told the bilingual guard, using the term for a decendant of the prophet Mohammad that was used to describe anyone on the prison staff. "I'm his new brother."

The guard led us into the new building. Like nearly every part of Evin that I was ever allowed to see, it was too bright. There were four doors on the left side of a long hallway. All except the second one were closed. Near the main door there was a step up that led to a kitchenette that was equipped with some empty and dusty cabinets, a dorm-sized fridge, a sink, and a drying rack, but nothing to put in them.

At the end of the hallway, past all the closed doors, was a vinyl-covered door on the opposite side of the hallway. It was open slightly, but I had been conditioned by those last seven weeks not to touch it. The guard pushed it open and said, "We'll come and unlock it in the morning, but we close it every day around this time."

We looked out at a narrow backyard that was about twenty feet across. There were some trees and through them we could see patches of night sky. My new friend and I were bewildered but looked at each other and smiled approval at our improved digs.

We had both just come out of solitary confinement and because he didn't speak any Farsi he was scared. Fair enough. So was I, but it didn't take geniuses to know that this was a big step up.

When I took stock of him I wondered if I looked as bad as he did: frail, hunched over, atrophied. I guessed that I did.

He was missing the top half of his right index finger, but judging from the look of it and how he maneuvered that hand, this was something that had happened many years ago. It would be a long time before I asked him how he'd lost it.

They'd let me out of solitary because they were changing tactics. I couldn't see enough to know why yet, but it was obvious.

By then the rounds of nuclear negotiations were happening more regularly. My brother, the *Washington Post,* and the legal team they had hired were working around the clock to try to bring me home.

Once I was out of solitary my world felt so big. I could hear the freeway nearby, and the mixed-gender laughter from Evin Darake, a small neighborhood of restaurants by a pleasant stream just beyond the prison wall. Yegi and I had been there with friends at a popular place called SPU the night before our arrest.

As the days in our new quarters began to pile up, I woke in the middle of the night sometimes, unable to get back to sleep. Somewhere in the distance I could hear a loud wordless call, and then moments later from another direction an identical response. *What was that?*

I wondered. *Night guards? Some kind of game?* I never figured it out.

Massive sycamore trees shaded the entire area around our walled-in compound. Their leaves began to fall soon after we moved in—it was fall—giving us a new task. "Keeping the yard clean is one of the privileges of living in a suite," we were told. We didn't care, because it gave us something to do.

But every privilege comes with a price.

I should have known that nothing comes for free on September 11. I continued to be terrorized.

The day after my move I was taken to the place I dreaded most, the big room where I had been brought on the first night. And I was blindfolded. I knew that meant another dead-end encounter with the Great Judge.

"You have not cooperated, Jason, this is a problem for you. You will go to court soon and you will receive a twenty-five-year sentence. This is the minimum. Others believe it will be more. Kazem, what is your opinion?" He was speaking from across the room directly in front of me.

"Execution," Kazem said calmly, sitting next to me.

"You do have a chance to go home, but it depends on you." He paused. "How is it being out of solitary?

Are you more comfortable?" He didn't wait for me to answer. "I brought you out of there and I can put you back. But I want you to go home. Your value, though, is very low right now and that is a problem for me. No one seems to care. Which is very surprising. Not the United States government. Not the *Washington Post.* Not even your family. You are forgotten. But we want to bring you back."

By then my mom and brother had made televised appeals for my release, one of them alongside Anthony Bourdain, who was on board from the first days after our arrest. The *Washington Post* was working frantically, in public and behind the scenes, to secure our freedom. I was already a "point of negotiation" in the now-regular talks being held about Iran's nuclear program, according to Secretary of State John Kerry. But I couldn't see any of that.

After more than fifty days, this was a clear acknowledgment from my captors that I was being held as leverage.

"You will do something for me and I will do something for you."

I listened, knowing I was in a bad spot.

"To raise your value I need to film you. To show them that you are alive and that we know you are a U.S. spy."

"But we all know that that is not true. And America knows that better than anyone."

"We know that you have been tricked. That you didn't know of your crimes. But, Jason, they are still your crimes. Even if you were being controlled from above." I had no idea what that meant, but it sounded so dumb.

"Whatever. So what do I have to do?"

"We will buy you some new clothes, because thanks to God you have lost some weight," he began. That was true.

"I'm pretty sure that didn't have anything to do with God. I thank *you* for that." Everyone laughed.

"And you would like to see your wife more. I can make that possible."

I knew he could but doubted he would.

"We need her to feel better before you go home. We are responsible for your health."

"So now what do I do?" I asked.

"Whatever we tell you."

"And once I finish with that I will leave?"

"God is great." Never the answer you want to hear from these guys. It means either "I don't know" or "no" and without the benefit of seeing into his eyes I couldn't guess which one it was.

Whether this was the plan from the beginning or something they'd come up with based on the bigger-than-expected international public outcry, I was now part of Iran's long history of American hostages and I knew it.

Late the next afternoon Kazem returned, and it was as though he was a different person entirely. His demeanor and facial expressions were completely different. Friendly.

Several days later the door to my new home opened.

"Come on, J," Kazem said in his Wanda voice, "we're going out."

He led me back to a room behind the doctor's office where there was a small area curtained off. He instructed me to take off my blindfold once I was on the other side of the partition.

I was in a changing room, and on a small rack various pieces of clothing hung, including my own. *At least I'm not the only one,* I thought.

On the floor were shoes, and I recognized the pair of brown slip-on Eccos that I wore the night of my arrest, amid the others. I had struggled to get them on my feet before because they were brand-new and hadn't loosened yet. Now, like my clothes, they had plenty of wiggle room. I knew I looked ridiculous in this outfit that I had to hold tightly to keep it from falling off me, even with my belt on. One of my socks was missing and they gave me a black one to wear with my own brown one.

But I was in my own clothes for the first time since being taken and I felt sort of normal, which is probably why I forgot to put my blindfold on as I exited the dressing room.

"Na, na, na!" A chorus broke out. "Blindfold."

"Sorry. Okay, okay." It was an awkward situation.

"Come on," Kazem said, leading me out of the building toward the concrete path that was the way out. We got into a car, which I could tell was more compact than the vans I had been transported in thus far. Maybe that was a good sign.

"Where are we going?" I asked.

"Shopping," Kazem said. "The Great Judge is a man of his word."

It was almost two months into my ordeal. The whole thing was so bizarre already, but this was taking it to new heights.

This time they sat me in the back of a passenger car, a black Peugeot 206, one of the most popular car models in Tehran. Yegi and her sister shared one for years before we got married. The "206," as they were universally known, was considered a cut above the Iranian-made Peykan and the Kia Pride, which long dominated Iran's roads. They sat me in the middle, sandwiched between Kazem and a guy I didn't know. Driving was the usual driver, and riding shotgun was a good-looking guy named Ali who obviously lifted a lot of weights.

A few minutes after we left the prison I was allowed to remove my blindfold—an Evin rule—and I recognized that we were in Pasdaran, one of Tehran's upscale neighborhoods, a part of town I assumed I was more familiar with than my chaperones.

"What am I supposed to do if someone recognizes me?" I asked.

"Act normal," Kazem said.

"And when they ask me about prison?"

"Tell them it's not true."

He was showing his ignorance of how reality works. I just nodded.

We pulled up to a branch of Hacoupian, a well-known Tehran men's clothing chain. It looked as though they had called ahead, because the staff didn't blink at the sight of me and eight chaperones, half of them in surgical masks, browsing the store.

"Stay away from anyone who looks like they could recognize you, and if you see anyone paying too much attention to you, tell one of us," Kazem commanded. That sounded so ineffective, but I said, "Okay."

The team naturally gravitated to what they expected—based probably on their own fantasies of taste—I would wear. Colorful and shiny suits, red shirts, a variety of bad ties.

"I don't wear ties," I made clear.

"But you're American." They were shocked. It was the same sense of wonder that Yegi's uncle had expressed when I decided to go tie-less at our wedding. "All foreigners wear ties at weddings. I've seen it in many films," her uncle said with authority. "This isn't a movie," I told him. He was upset, but when I reminded him that I was paying for the whole event he quickly got over it.

I walked around the store and picked out a couple of shirts and three pairs of pants, finally settling on a blue shirt with small white checks with the kind of slim cut that had never before been an option for me. The pants were light brown and too long.

"If we send them to the tailor it will take three days," the shop clerk told us at checkout.

The team deliberated. *Who will come back to get the pants? Do they have the budget to pay for tailoring? Is it even necessary?*

"We have our own tailor," Siamak finally said, handing over a debit card.

"Password?" the clerk asked, because in Iran you actually give the person running the card your PIN.

Even these guys, who were supposed to be the nation's last line of security.

"Wait, he needs a belt," Kazem remembered. That was true.

They picked out a belt with an "H," for "Hacoupian." It was very appropriate, because it was the sort of accessory I would never think of picking out for myself.

We left the shop, my entourage and me, and they said we could take a little walk. We went up the street a couple hundred feet. It was the most I'd walked without being forced by a wall to turn in more than two months. It felt great to walk in the autumn air, in real shoes, in straight lines, even if I was being escorted by the same guys who had taken me captive and happened to be packing heat.

My life had taken so many twists already, I liked to believe that this was just one more of them, and it would soon be behind me. A good story to tell that would remind my friends back home that I was on a different kind of trajectory—one of them, but not really.

We stopped in front of a florist.

"Let's go in," Kazem said.

He can't be serious, I thought. As he walked in I realized, *He is.*

"The Great Judge said you must buy flowers for your wife," Kazem said. He had no idea how creepy that sounded.

"I don't have any money," I told him.

"We will put it on your account," he promised.

We went down a flight of stairs and entered the basement-level shop. In the five years I'd lived in Tehran I'd bought more flowers than I had in my life-time in America. Flowers and pastries. You don't visit an Iranian's home, especially for the first time, with-out one or the other. My guards gravitated toward the cheap stuff, and I went straight for the more expensive ones. "My wife likes tropical flowers," I told them.

The florist put together a bouquet, handing me each flower to smell before he added it to the bunch. He was a gentle guy, in his fifties, sort of chatty. He wanted to talk about love and its power. I wasn't in the mood to make small talk about subjects I understood but couldn't touch right then, but I agreed with everything he said.

"Your heart is big," he said to me. "I wish you and your wife many years of happiness and many children. They are life's sweetener."

"Yes, so I'm told." We returned to street level, and a popular confectioner's was right there.

"You must buy pastries for your wife."

These dudes are unbelievable.

I knew it was an opportunity, though. She was starving and so was I. And so was my cellmate.

Again, they started choosing based on their tastes—a bunch of cakes filled with fruit jelly—before I stopped them. "My wife doesn't like those, and neither do I."

"Excuse us. Please choose what you like."

I had the clerk fill a massive box. Two kilos; a completely normal and perhaps even small purchase by local standards, but much more than Yegi and I would be able to eat ourselves before they went bad. My stomach ached just looking at the sweetened cream atop the individual cakes, some of them filled with nuts, or chocolate, or simply more sweet cream.

And that was the end of the fairy-tale adventure.

"Let's go, J," Kazem said, pointing the way back to the 206.

We drove back to Evin on the highway, and when we got close I was instructed to put my blindfold back on. We wound through the prison compound, until the car stopped and I was told to stay put.

Someone got in.

"Salaam, J." It was Borzou. I'd heard that voice twice before over the previous few days. Kazem had said his arrival meant that my story was coming to an end. He was the closer. I had no idea if that was true. "You are finally cooperating. We appreciate that. You'll do what we say and hopefully, God willing, things will work out for you and your wife. You will say everything on camera."

"What everything?"

"Everything we tell you to say. Don't disappoint us."

I was beyond fear by then and had moved on to hopelessness. I didn't answer.

"Go enjoy some time with you wife. You do for us and we do for you. We aren't false promisors." And that was it.

Siamak led me through the short corridors of the prison offices and through a side door. I was greeted by a guard who used a metal-detecting wand to pat me down. We stepped through a door—I was outside again—and then I was led around the side of the building to another door.

"Take your shoes off," Siamak told me.

I entered a room with a machine-made rug on the floor and a wooden coffee table surrounded by four green vinyl-covered chairs. In the corner was a smaller table with a telephone on it. The air conditioner was turned on low, so the temperature was pleasant. He placed the box of sweets on the table and handed me the bouquet.

Within moments I heard my wife's voice and footsteps. She entered the room and began to cry. It took me a moment to be sure, but they were tears of happiness.

"What are these new clothes? Are they letting you go?"

"They took me shopping today," was all I could answer.

Siamak left and we embraced for the first time since our capture. He called through the wall, "Please keep your hijab on and do not use this room for husband-and-wife activities. You are on camera." We sat and held hands. I put an arm around her.

Yegi and I tried to catch up, but although we'd been away from each other except for a couple of brief and heavily monitored encounters, there wasn't much new we had to share.

"Why did they take you shopping?" she asked.

"Taped confessions," I told her.

"You're going to do them?"

"I don't have a choice."

"Why are they doing this to us? We didn't do anything."

"I try not to ask myself that question."

There was a knock on the door. It was Siamak. Our bizarro day kept getting weirder. He was carrying a large silver serving tray. On it were two small plates, each with a chicken leg quarter; another one with two pickle slices; and a small stack of the same paper-thin lavash bread they gave us for breakfast. He left us alone. It was exactly the same food we would have been given in our cells, but it was on plates.

Yegi ignored the food, just wanting to talk. We reminisced about a thousand experiences we had had in our five years together. The present circumstances, though, kept getting in the way.

"Why?" She wanted to know was this happening to us.

"I don't know, baby, but someday we're going to laugh about this. I promise. For now eat something."

"I can't. I'm not hungry," she said.

"Really?" I said, and opened the box to show off the cakes.

"Oh my God. Those are ours?" she asked.

"Yes."

"Who's going to eat all of that?" It truly was a lot of sugar for a couple of starving inmates.

"We are. Have one." I put the box in front of her.

"So many choices." It was true. We hadn't been able to choose anything for the past two months.

We bit into the creamy sweets, a momentary escape from purgatory. Bliss. Life still existed. It was a tiny but important reminder that we had plenty to look forward to.

Moments later there was a knock at the door again.

"Miss Salehi." It was Siamak. "It's time to go."

She left and I changed back into my prison clothes. I grabbed the nearly full box and put my blindfold on my forehead. I was going home with treasure. Not the one I wanted, but the only one I had then. How was I going to explain this to my cellmate, a guy I communicated with through a combination of single-word sentences and male grunts?

A guard led me back, past the airport-security-style wand-er. I entered our compound and called out, "Mirsani. Sweets," handing him the box.

His eyes got very wide. It was about nine P.M. He was lying in our bedroom, watching TV.

They had a full script for my confession—an adaptation, really, of their translated and abridged version of my interrogations.

The case that they were trying to make against me was that, as a member of the American press writing what could only be perceived as neutral stories about Iran, I was attempting to soften American public opinion toward the Islamic Republic.

My "mission," or "mizhan," according to Kazem,

was that by improving this image America would some-how infiltrate the Iranian system, fill the halls of power of the Islamic Republic with like-minded Iranians, and worse yet Iranian dual nationals, in the process gutting Iran of its revolutionary ideals. That part was ridicu-lous. But, since I understood completely that I had lost before I started, this seemed like a much better narra-tive to work against than the one that they had wanted me to cop to: that I was a CIA agent.

On the one hand it meant that I had an endless list of accomplices. Journalists, scholars, businessmen, westernized youth, unemployed dreamers, tech-savvy teenagers, and Iran's current presidential administra-tion and cabinet.

I would not be taken down for nothing all by myself, and if they wanted to they would have to go after count-less others, too.

This wasn't a strategy, but a way of defending myself when I had no other resources. Shift the blame, but in a way that it doesn't hurt specific individuals.

The Islamic Republic was born on the premise that America is the Great Satan, but that enemy's menace is distant and hard to touch. Even when sanctions were at their height, there weren't American bombs being rained on Tehran or other cities.

Saddam Hussein handed the ayatollahs a gift when

he invaded Iran in 1980, setting off an eight-year war that the regime still leans on to justify its revolutionary rhetoric. The blood of hundreds of thousands of young soldiers and civilians was a small price to pay for Khomeini to secure the system's reach for a generation.

The end of that war coincided with the purge of thousands of regime critics, mostly those accused of having ties to the MEK and Sunni Kurds, which served the purpose of solidifying support for the system or at least cultivating deep fear of it during peacetime.

In the late nineties the new evil was represented by President Khatami and his reform movement. As a mullah and a product of the system, he was easily brought into line, politically neutered in the process.

The specter that Mahmoud Ahmadinejad cast on the Iranian system and society was harder to combat. He had been the hard-liners' darling and they doubled down on him in the wake of his highly suspect 2009 reelection, only to be betrayed by him and his inner circle, who saw themselves as bigger than the institutions of the Islamic Republic and its supreme leader, Ayatollah Ali Khamenei.

Ahmadinejad and his cronies had emptied the country's coffers, squandering in eight years oil revenues that amounted to more than all of the country's petroleum income combined up until they took power.

Their corruption was astounding even to Iranians well versed in living under rotten leaders, something they've known intimately for centuries.

When Hassan Rouhani was elected it was the promise of the removal of sanctions and a renewed cooperation with the rest of the world that swept him to a dominating victory. A return from isolation, it was clear, was what most Iranians wanted, not revolutionary slogans, material support for Palestinians, or a minimizing of foreign—mostly Western—influence in their lives. If anything it was more of the latter—as much as they could get their hands on—that most people craved.

The mystique of its modern delights was the secret weapon America had used to seduce Iran, all the way to its top elected officials. International commerce, technology, foreign brands, and entertainment. Fun. These were the Islamic Republic's true enemies and they had infiltrated every layer of the society.

I pointed out to my captors when they would air their ridiculous conspiracy theories that, "Even the supreme leader has a Twitter account," although it wasn't verified, as Rouhani and Zarif's were.

This was the premise, or "plot," that they had been dancing around for weeks but were too dumb to name. I fashioned it into a narrative for them. A true one that made perfect sense: it worked with their silly ideological worldview and was viewed, rightly, as completely outlandish by the entire world.

Writing it wasn't difficult. Taking bits of floating information, tying it together with quotes and figures, and packaging it as a report on a trend, idea, or news. Isn't that what we do? Helping the public make sense of nonsense?

It was an all-in gamble, but what did I have to lose? My wife and I were in prison. My mother and brother were on the other side of the world. I was careful not to implicate anyone else in any sort of actual wrongdoing or crime.

It was, in reality, the only idea I was able to come up with in my tiny vacuum-sealed world of solitary confinement and I stuck with it. The semi-insane machinations of a man being pushed toward lunacy.

I took great care in what I implied and stated clearly, "This is not an elaborate plot to undermine the Islamic Republic, but rather the direction that the world is moving. Ideology is over."

What did I care? I was ruminating on massive questions of geopolitical and historical significance with an audience that had already decided I was public enemy number one.

"J, whether you say it or not, you're guilty, so better just to say it," Borzou reminded me.

Of doing what exactly, though, none of us were sure.

So I tried to hedge my bets. "I'm not an agent of anyone except the *Washington Post*," I wrote. "I'm not the problem. I didn't start it. I'm just tracking this phenomenon."

"What do you think of public diplomacy?" Borzou asked me in the middle of the forced confession tapings.

"Why does it matter what I think? Engagement between individuals in different fields is an old practice that countries, including America and Iran, encourage to increase knowledge, share expertise, and reduce tension. It's a policy that both Tehran and Washington are promoting right now."

He shook his head and chuckled and then handed me a piece of paper. It was a printout from the State Department's website.

As I glanced at it he read a Farsi translation of the same, but in a sinister-sounding tone.

"The mission of American public diplomacy is to support the achievement of U.S. foreign policy goals and objectives, advance national interests, and enhance national security by informing and influencing foreign publics and by expanding and strengthening the relationship between the people and government of the United States and citizens of the rest of the world."

When he finished he stared at me over his surgical

mask and paused for effect, as if I had been caught in the act.

"J, if this is not 'spionage, what is it?"

"What are you asking me for? All I know is that it's perfectly legal. Anyone invited to Iran to take part in public diplomacy programs has been vetted before they come."

But I was starting to see what was happening. They were trying to make the case that I was somehow the architect of a decades-old policy of outreach that had yielded such negative results for their plans that the IRGC felt they must shut it down.

In reality I had explored several projects, even writing proposals to bring Iranian journalists, including hardline ones, to the U.S. to do reports on positive inroads their countrymen had been making in American society. Another one would have brought Iranian tech students and young entrepreneurs to Silicon Valley to take part in startup seminars and provide them tools, training, and contacts they could bring home with them to Iran. Good projects.

Throughout the interrogations, I was completely honest about these activities. Even if my captors could not see them to be innocuous, I thought, it would also be difficult to publicly spin them in a negative light. But of course they could. It was what they did.

"It really doesn't matter, though. It's your hidden diplomacy that we're concerned with," Borzou explained solemnly.

"My what?" I asked, almost certain they were inventing a new term.

"We know that it's your mission to make the negotiations succeed," Borzou said matter-of-factly. "Now you just have to help us prove it."

I stopped to ponder how stupid that sounded even in the vacuum.

I was completely exhausted by the end of my completely optional forced confession tapings. They had gone on for days, with retakes and a lot of bargaining. Kazem pushed to get me to say things that weren't true; I resisted by saying things that didn't implicate anyone in actual crimes. But who was I trying to fool? All this was just putting nails in my own coffin to speed up the arrival of the imaginary process of Iran and the U.S. negotiating over me. It was a Hail Mary pass from fourth and long. What I didn't know was just how long it would take for the sides to return to huddle from their time-out.

But I got my jabs in. I repeatedly told my captors that, save for beheading me, there was absolutely no difference between their behavior and that of the Islamic State. They hated the comparison, but they knew it was apt. "Why can't you just admit that I'm a hostage and nothing else? Say the word."

They wouldn't do it.

But everything leading up to the confessions and all that happened after was in line with the behavior of captors negotiating to free a hostage. "We have to do this to raise your value. You're worthless right now. No one cares," they kept telling me.

They made me read a letter to Obama in which I had to apologize to the people of Iran for contributing to their suffering caused by years of sanctions, say that I intended to overthrow the regime in Tehran by helping cultivate human bonds between our people, and ask him to do more to win my release.

It was humiliating, but not any more than any of the rest of it. I was tired, and I didn't care. I just wanted my life back. But they just wanted me to keep performing.

"J, when we invited some of your friends here to ask them about you"—he meant interrogate them—"they told us you are a very good singer."

"That's not true. I *like* to sing, but I'm not a good singer."

"You must sing for us."

"Goddamn you." Oddly, that's not considered blasphemous in Iran. "All of you. I've acted enough for you already this week."

"Come on. Just one. We're all colleagues here," Borzou said, repeating his catchphrase meant to imply that he and I were both covert agents.

I thought about my options. I had no idea whether my release might be imminent or very far off. I'd already folded and done the one thing they wanted me to do. I was as close as I'd ever been to saying "uncle."

"Okay," I told them, "but you have to stand up."

They rose. All of them.

And I launched into a gloriously lonely rendition of "The Star-Spangled Banner."

When I finished they clapped.

"What was that?" Borzou asked. "It was very lovely, especially when your voice cracked. Very emotional. So much feeling." He was messing with me.

"That was the American national anthem," I told them. "You guys better be careful, you just betrayed the Islamic Republic and paid the ultimate respect to the Great Satan. This is a very big crime."

Of course, they were untouchable, or at least they were sure they were. Everyone laughed for a minute, including me. And then they led me back to my cell.

I spent the next several days wondering when they would broadcast what they considered my confessions and how bad they'd look. One late morning I was brought to the room where Yegi and I had enjoyed our dinner and pastries after my shopping trip several weeks earlier. I hoped she would be joining me there. When she walked in wearing her own clothes my heart sang, but then it sank. She had been released; that was very good news. I had not.

Once she'd started to acclimate to the new reality of being the wife of a convict, she set out to get me out, but also to do whatever she could to make me more comfortable. Understanding the conditions of prison, she knew well what I was missing.

It took weeks to get permission, but after much jostling she was able to bring me an initial care package. Home-cooked meals were off the table, but they allowed a small French press coffeemaker. She brought along two factory-sealed cans of Illy espresso, which we'd bought on our last trip to Dubai. Initially they said it was contraband.

"How do we know this is what you say it is?" the deputy warden asked me.

"Open it up and I'll make you a cup."

"But if you get sick it will be our responsibility. You will have to sign saying we are not responsible if you are poisoned."

It was obvious there was no such document available to sign.

"Give it to me. I'll sign it right now." I wanted coffee that badly. It had been almost three months since my last cup, and the withdrawal headaches had passed unnoticed amid all the trauma of the first days in solitary. I knew access to a basic comfort of home would go a long way.

After several days they relented, and I had that small privilege.

Yegi also was able to bring me some books. We had a decent library in our home office, titles that I had collected over years of short visits back to the U.S., but her choices didn't reflect the range of our collection. She brought *The Power of Now, The Alchemist,* and several others by Paulo Coelho that I knew I wouldn't read.

"Thanks for these, but try to bring me a few more."

"Like what?" She was trying to lift my spirits and transport me beyond the walls through self-help literature, and all I wanted was to feel some camaraderie with the people of the past and the injustices they survived.

"Bring me whatever Orwell we have at home and any history."

She brought me some of my own clothes, too: a silver Adidas sweat suit, a pair of green New Balance sneakers, and some T-shirts, including one from her own collection: an oversized airbrushed one of Leonardo DiCaprio in *Titanic*.

Drinking a cup of coffee and wearing my own clothes, I realized, quietly, that these slivers of normality could help me fool myself into being a little less in prison. That might become essential to surviving.

Efforts to spring me from Evin had begun as soon as it was clear that Yegi and I had been arrested. By the time I met Kazem for the first time the *Washington Post* had hired Bob Kimmitt—a lawyer at a DC firm called WilmerHale who had conducted negotiations with Iranians in the past and worked to bring home others detained there, including Maziar Bahari—to represent it in their quest to get my freedom. My brother had engaged with Iran experts, some of them self-proclaimed and others with actual bona fides, and had taken the most important step of having my Gmail account suspended. The damage there had already been done, but likely in the most rudimentary of ways, and it became clear to me, within the first weeks, that my captors had had very limited access to my communications; it started and then abruptly ended. If this was all they were working off of, I thought, then this couldn't go on for very long.

The debate inside my family about whether or not my mom should travel to Iran began to heat up once

Yegi was released from prison in early October 2014. She was adamant that my mom should come immediately, that her presence would be a net positive and that the authorities in Iran were less likely to do more lasting harm to me with an American mother—who happened to be an Iranian citizen, and therefore had the right to come and go as she pleased—in town.

Ali's concern was that if Mom came to Iran she wouldn't be allowed to leave. Yegi and I, independently and later together, thought that it was highly unlikely that the Iranian authorities would give her any problems, but we understood the concern and their logic.

With my mom living in Istanbul at the time, my brother in Marin County, my employers in Washington, and Yegi and I being held captive in Tehran, the challenge was to get everyone reading from the same playbook.

Everyone wanted the same thing: to get me out of prison. But the path to that looked significantly different from each individual vantage point.

Once she was set free, I reminded Yegi of our shared goal, and assured her that my mom and brother would be doing everything within the bounds of their power and what they thought was right to help win our release. I was sure of it.

7

Life in Prison with Mirsani and a TV

It turns out Mirsani lost his finger as an infant, sticking it into a machine that he shouldn't have. After many months together I still never learned what kind of machine it was, which should tell you how limited our communication was.

What I did know was that he was forty-one years old. He was from Jolfa, a city in the Republic of Azerbaijan that borders an Iranian city of the same name. He imported consumer goods from Iran, where they were cheaper. Candy bars, rice, and diapers, which everyone in that part of the world calls "Pampers."

I tried to teach him some English and as you do if you're not a teacher you start with the things around you.

"Star," I said, pointing to the night sky.

"Star."

"Good. Clouds."

"Clowvs."

"Pretty good. Moon."

"Myoon."

"Moooon."

"Myooooon."

He taught me Azeri words as well, and I assume I sounded just as funny to him.

He had some sort of problem with his ears that affected his hearing. Azeris, or "Turks" as they are referred to in Iran, often speak very loud and because of his hearing issues Mirsani spoke even louder. I never knew whether he was angry or just animated, but in all the time we were together we never fought. Not once.

Part of that was probably because we had a TV in our cell. The great pacifier. He could watch TV more than anyone I've ever met, which is incredible since he never understood completely what was happening on the screen.

I tried to stay in shape by walking laps—hundreds, I called them, because I did them in sets—in our concrete yard, and once they allowed me books I poured myself into those. The best mental escape.

But Mirsani didn't read. "Bad for my eyes," I understood he was telling me.

We learned a lot about each other without really having a common language. He had a wife and three kids. The oldest was already in her twenties. Mirsani had been a sniper in the Azerbaijani army during its war with neighboring Armenia, which I could never wrap my head around, because he was right handed and missing the essential part of that trigger finger. He sniped lefty, it turned out. I'm still not sure how Azerbaijan did in that war.

We had already figured out that we had one very important trait in common: our appetite, which had atrophied along with our muscles while in solitary.

In prison, especially if you have no idea how long you'll be there, your aim isn't to kill time, it's to conquer it. Days naturally turn into segments. Routine—never my strong suit—had become everything. No surprises in the day, thank you very much.

As on the outside, the best time waster is a television, but what we had from the moment we stepped out of solitary wasn't just any kind of TV; it was the full package of Iran's state broadcaster in all its propagandistic, idiosyncratic, copyright-skirting, America-hating, America-obsessed, anti-Semitic, self-unaware, nationalistic, narcissistic, sexist, Shia glory.

It was horrible, but we quickly learned to be conscious consumers of the tripe. We had to, because it was our only source of information and diversion, and from the moment Mirsani woke up in the morning, it was turned on.

In those early days we filled ourselves on movies. The night we moved into our new room we quickly worked through the available channels. We came upon the early moments of *Enter the Dragon* dubbed into Farsi. I knew what I was looking at and so did Mirsani. "Ahhh. Bruce Lee," he said. "Good."

We settled in to watch. There was an advertisement at the end of the film notifying us that it was Bruce Lee Week. After the despair of solitary it was the first thing we had to look forward to and we clung to it.

Next it was Jackie Chan Week, and I can report that I've now seen more Jackie Chan movies than I previously knew existed. Jackie Chan Week actually became Jackie Chan Month, and because many of his older films are short, they would often play two of them, back to back. We watched so much Jackie Chan that we started to joke that he was our third cellmate. And when Jackie Chan Month was over he would pop up for a visit often in the distinctive voice of the dubbing artist who did his movies.

A character would speak—Johnny Depp or Ryan Reynolds—and I would say with confidence, "Jackie."

Mirsani would lean closer to the TV from his pile

of blankets, because that was our only furniture, and listen for a moment. "Haa," he'd confirm. "Jackie."

We heard that voice a lot.

But more important than the entertainment value was the window on the world that the television provided. It was the first actual evidence I had that my captors were lying to me about everything.

They said the *Washington Post* was doing absolutely nothing on my behalf. Not even talking about me. I found it hard to believe, but what did I know? I was not a permanent employee of the *Post* then. I was, like most foreign correspondents today, a contractor, a permanently hired gun. But any fears I had of that detail weakening their resolve to work on my behalf disappeared on September 24, as I watched, with a half-open jaw, an exchange between Iran's president Hassan Rouhani and the *Post*'s executive editor, Marty Baron, being replayed on Iran's state television. They were talking about me.

In a conference room filled with reporters, some of whom I recognized, others whom I knew personally, my hope was validated.

"First of all I just want to take this opportunity to revisit the subject of our correspondent Jason Rezaian and his wife, Yeganeh. I would like to say to you personally that we believe that he deserves his freedom and we ask the government of Iran to release him," Marty began. "But I want to ask you how the Iranian government can justify imprisoning a good journalist. I think you know he's a good journalist and a good person. And having him imprisoned for two months and interrogated for two months, how is that possible?"

Much of that initial hope, though, was immediately dashed with Rouhani's smug and evasive response. To my knowledge they were his first public words about my case, and as he has to this day, Rouhani refused to say my name.

"I myself don't have any judgment about a person being investigated by our judiciary."

Years later Marty's straightforward question remains unanswered.

Before prison, I never watched Iran's local TV, but now it was my only option. Through consistent watching I gained insights into the regime I probably never would have otherwise, despite all the censorship, or maybe because of it.

What was considered interesting, what was omitted, and what had to be shown to get people to tune in to a domestic broadcast rather than a foreign-based satellite one? I was surprised—sometimes for depressing reasons—on a daily basis.

Every night at eleven P.M. a channel called Namayesh—loosely translated as "Showtime"—would play a Hollywood movie. It didn't take long to recognize that the films were chosen for their themes portraying American power, and Western lifestyles generally, in a negative light.

It was the quintessential Islamic Republic trade-off: give the people what they actually want, but try to spoon-feed them an extra helping of ideology and rhetoric.

In addition to their choices, the handiwork of the state media regulators was one aspect of watching a movie that never got old. In a society where sex out of wedlock was not only a sin but a crime, not to mention the even more evil acts of homosexuality and alcohol consumption, I imagined the decision to show Hollywood films created job security for legions of censors.

And translators.

You want examples? There is no dating in Iranian films. No boyfriends and girlfriends. Any unwed people who have contact with one another are "fiancées." When characters in a movie walk into a bar and order a "cold beer" in Iran it becomes a "cold soft drink."

This can go to extreme lengths and means that a lot of films are cut into tiny versions of themselves. When

Yegi and I were allowed to see each other one day I told her that I'd watched *The Big Lebowski* the night before. She was stunned, because we had watched that uncensored at home.

"How long was it?" she asked.

"About fifty minutes."

No white Russians, no marijuana smoking, no urinating, no Bunny Lebowski. Just a couple of guys bowling and a Persian rug.

The Untouchables is another one that sticks out. The whole story line of bootlegging and prohibition was eliminated, turning Al Capone into little more than a murderous tax evader.

In *Philadelphia,* Tom Hanks and Antonio Banderas's characters were brothers.

In that cell, though, there were also movies that did exactly what I needed them to when they came on the screen. I'm thinking about *Moneyball,* which, on the surface, is an odd choice for Iranian television.

True, my last Tehran-bylined article before my arrest was an in-depth look at Iran's budding baseball community—further proof of my supposed goal to usher American values into Iranian life—but let's be serious, baseball wasn't taking off to the extent that people wanted to watch movies about it. No, this was all about a love for Brad Pitt.

It was the first time I had seen one of his movies on Iranian TV and I was pleasantly surprised that he was one of the actors that got dubbed by the Jackie Chan guy. Billy Beane and Jackie Chan are forever linked in my heart.

How could I not breathe out a huge sigh while watching the recounting of an epic baseball story—the Oakland A's twenty straight wins in 2002—while I sat in Iranian prison? *If this can make it all the way to me, here, these bastards can never win.*

I only wish people around the world could wrap their head around the fact that Iran, or the Islamic State, or fascists, or the Soviets, never once posed a threat to our way of life and probably never will. We have the ultimate soft weapon, Hollywood, and it's one that no matter how hard they try, they can't replicate. They don't know how to be all things to all people the way we do. You can't learn that; you either have it or you don't.

But here I was revisiting one of my favorite sports moments—a game that I attended with my dad and a couple of high school friends—as it was dramatized and played back to me in my prison cell more than seven thousand miles away from where it happened, in the Coliseum in Oakland, over a decade later.

There was no one I could explain the significance of

this to, but that didn't matter. Hollywood touched me, just like it was touching people all over town. You can't undo that with rhetoric, especially not if it's about guys that died fourteen hundred years ago. As usual, Iran was fighting a losing battle.

Every time I would point this out to Kazem his response was, "Yes, but we have Islam," and he meant it, or at least he thought he did.

He knew I was right, though. If he loved Will Smith so much that he wanted to be played by him in a movie that would probably never get made, what about the many millions of young Iranians who were already openly opposed to the regime and everything that came along with Islamic rule?

American culture could not be shut out. I felt it to my core one afternoon, when alone in the cell something incredible happened: the Lucasfilm logo came on the TV, and boom, the Star Wars crawl started. It was *Episode IV: A New Hope.* I made the jump to light speed—and in Farsi—back to the hyperspace of my childhood.

Mirsani, after taking in the first few moments, turned over and put his blindfold back on to nap. "This film is kids film," he said, his way of saying a movie was too unbelievable to maintain his interest. Instead he gravitated toward more highbrow material.

He loved Mr. Bean, whom he, like the rest of the world, knew by name.

He loved violence and big guns. And he had seen plenty of movies. Sylvester Stallone was Rambo and Dolph Lundgren was Ivan Drago, the name of his Russian character in the later Cold War–era classic *Rocky IV*, in which he costarred opposite Rambo.

Citizens of former Soviet countries don't really know much about the twentieth century. Same with present-day Iranians. Their version of events is a strange hybrid of the nationalist bits of their own official history spliced with the shinier Hollywood one.

Mirsani loved anything with Bruce Willis, but not as much as he loved Jason Statham; if there was ever a movie made about my ordeal, he reported, that's who he wanted to play him in it. This was becoming a running theme.

We would laugh at particularly violent scenes, especially if they seemed excessively fake.

And watching prison movies in prison? We knew no greater joy. Any time solitary was mentioned we would feign sympathy. "One week? Awwww."

We watched and we watched, and thankfully the television was never taken away from us.

I watched the news religiously. Propaganda media, when viewed critically, can be an incredibly revealing window into what a state apparatus is trying to accomplish. And when I realized that my fate was absolutely intertwined with the fears of a major part of the ruling establishment over the impending nuclear deal and new opening with the West, I started to wonder if they would ever be able to undo what they had started with my arrest.

In typical Iranian fashion their response to the increasingly incredulous reactions of the world to my arrest was to double down.

I felt like I was losing my mind trying to make sense of it all, and then one day as I watched the news on Iran's state TV, they were doing a human interest story about a guy from Ohio (so much of Iran's state television is connected to, and often produced in, the Great Satan). He wanted to raise $10 on Kickstarter—*See, it exists!*—to make *potato salad,* and somehow, for some reason, he ended up getting $55,000 pledged to his project. So he threw a party and raised a bunch of money for charity. It was quintessentially American.

Sitting in Evin Prison, very far away from Ohio, from my wife, my mother, my brother, my colleagues, and my home, I felt a little envious. A little bitter. Not just because guacamole is so much better than potato salad. I could do little more than shake my head.

Sometimes I just threw myself into infomercials.

I was struck by the fact that Iran had finally picked up on the concept. It seemed so out of place. I hadn't seen one in many years. Or maybe it was because they were exactly the same in tone as the ones I remembered from the early nineties, selling the same crap. At this very moment, all over Iran, BluBlockers and the Club are enjoying a second life.

I noted it in an interrogation one day, when it was becoming clearer I wasn't going anywhere and that nothing I said or did had any bearing on how long I might be stuck there.

"I understand you are afraid of America's influence. The cultural invasion is very real." I meant what I was saying but wasn't implying I had anything to do with it. "But it's too late. It started so long ago. Your advertisements are just like American ones. And do you know why? Because they work."

"Perhaps you're right," Kazem said, with something between indifference and lamentation.

"This isn't my fault. I am not an advertiser. I don't have any product to sell. I don't approve what goes on your TV. Those are decisions made within your precious system. I'm nobody."

"No, J. You are somebody. We are just not sure who yet."

I remembered a hundred days walking around Tehran

and all the signs of consumerism. It didn't even make sense to call it Western anymore, because it was everywhere. Freedom today, I would tell anyone who asked, is not the concept we have in America and Europe of choosing our leaders and expressing ourselves however we please. It's the freedom to buy stuff and to access entertainment. If you give Iranians—most people, actually—unfettered access to *that* while increasing their income, they will love you forever.

That is exactly what the IRGC feared. Sanctions didn't bother them until it hurt people's purchasing power and then it became a problem that needed to be rooted out.

How I was catapulted into being the architect of that problem is another issue. Some will say it was because I was an American, working for a high-profile American media company. But more than that it was that I was using that platform to tell these stories. Explaining parts of Iranian life that Iranians didn't see themselves. That really didn't take much, just a fresh set of eyes.

But I couldn't explain that to Kazem or Borzou. These guys were black and white. No-dimensional.

I told Borzou, "You are not very good at your job."

He was expecting a joke, because I worked those in sometimes unexpectedly. "Why not?" I could see his eyes smiling, though his face and nose remained covered by his surgical mask. He was waiting for a punch line.

"You are completely unable to put yourself in anyone else's shoes. That's why you fail. That's why the nuclear deal is happening even though you were sure your 'dear leader' would stop it."

He was angry, but he knew I was right.

"J, when you get out of here we want to stay in touch with you. Would that be okay?"

"I don't know. If America is as bad as you say it is and you guys are willing to lock me up for nothing, what will they do to me if they find out I'm talking to you?"

"We'll make it easy for you. We have people everywhere. We can even have meetings in Washington."

"And what would you want from me? What would I have to give you?"

"No, J, not like that. It would be win-win, as they say." That was Rouhani and Zarif's catchphrase for the nuclear negotiations. He knew I knew. "We would just like this kind of information you're giving us now."

"This isn't information, it's what's going on right in front of us. Just watching your censored TV anyone can see what an unjust and ridiculous system this is. And un-Islamic."

He didn't want to change the subject. "You just give us a report from time to time. Maybe not for several years, but give us a report that will help us understand what's happening. Or maybe we can give you a piece of information that you can pass on to your friends in America to help them make a decision." He had over-played his hand and now switched course and was baiting me.

"No, I don't think that's possible. I don't work for you and I don't work for them. I work for the *Washington Post*."

"What's the difference? It's all the same thing." He was frustrated. He only ever came to see me at the end of the day, when he must have been tired.

"Sure. I'll do exactly the same thing for you. I will make reports for you and publish them in the *Washington Post*. My work is completely public and transparent."

"Yes, that's why it was so hard for us to find you." Borzou nodded, remembering he was playing a part. "How will we know they are for us or if it's informa-tion we want?"

"You'll know," I assured him. "You'll know."

Most of the time I felt as though I were dealing with second graders. Second graders who admittedly had a lot of power over my fate.

With the TV on in the background, I was always waiting for a glimpse of people I knew. People from my real life. It wasn't hard to find them. I knew the entire international press corps in Tehran and they would make appearances on my screen often as would other friends and colleagues when Iranian officials made trips abroad.

But during a report on torture at Guantánamo I saw a familiar face from a different part of my past.

It was Christopher Hitchens. It was a short clip of his being water-boarded. He had submitted himself to that so he could report on the torturous practice for *Vanity Fair*. It was in 2008. I had seen it before. A lot of images popped onto my Iranian TV screen during those months, but that one kept coming back whenever the subject of American torture resurfaced.

The juxtaposition was hard to swallow.

One of the world's biggest critics of religion, dictatorship, and hypocrisy, and one of Salman Rushdie's best friends and protectors when Ayatollah Khomeini famously declared a fatwa against the author in 1989—so one of the Islamic Republic's ideological foes—posthumously became the face of Iranian propaganda's anti-Guantánamo material.

It was one more way that my old life, my American

one, reached me in Evin. Every time they aired that clip of Hitchens, and it was often, I was transported. I had no doubt that if he were still alive, he'd have been one of those fighting passionately for my freedom.

I don't think I need to tell you how incredible it was to be twenty-three and in New York City and one of the very few straight male students at Eugene Lang College, part of the New School, an extremely open-minded liberal arts college in the West Village.

I lived in the East Village on Sixth Street between First and Second Avenues. Yes, the block that has all the Indian joints on it, one of which I lived above, and I still kind of smell like curry.

No one in the nineties talked about the New School or its illustrious history, but in the classroom I was certain I had the best teachers possible for me.

I took a course in the history of reggae music. The instructor was Herbie Miller, who had a seemingly endless list of stories about the biggest names in Jamaican music. It was a one-credit class that met on Monday nights during a very cold winter.

In the last session of the semester, as we listened to some of reggae's essential songs and very old recordings of the Nyabinghi drumming that is at the heart of the music, another student asked Herbie why he was such an authority.

"Oh me, mon? I dabbled in da industry for a couple years," he replied coolly, explaining that he had been Peter Tosh's longtime manager and had traveled the world with some of reggae's biggest acts.

It was just like that there.

I had a class in jazz culture with the Pulitzer Prize–winning critic Margo Jefferson.

I wasn't there to study music, but I was trying new things and stretching out my horizons.

I had reconnected with an old friend from childhood, Noah, also a student at Lang. Because of my circuitous educational journey, which had taken me to four colleges—my dad used to call me a Road Scholar, apparently unaware that that's not how you spell it—I was a year behind Noah by the end of college.

He and I would meet up often for drinks. A few weeks before the fall 1999 semester he told me, "There's a class you have to take. I wish I was still in school just for this."

"What's the story?" I asked. Noah had never led me astray.

"It's a literature class"—honestly, not really my thing at the time—"but the professor is this guy who's a writer I really like. He absolutely skewered Bill Clinton, Mother Teresa, and the Dalai Lama."

As he was known to do, Christopher Hitchens shook my world.

The class was a comparative literature seminar that looked at U.S. and British authors from the nineteenth and twentieth centuries. The class met—after our first session in a fluorescent-lit windowless classroom—at Borgo Antico, an Italian joint around the corner on Thirteenth Street.

The deal was that everyone had to order something—either food or beverage—and that Hitchens bought the first round. Our assignments were to read works of our choosing by a specific author before the next week's class and be prepared to discuss them.

One evening Hitchens didn't name an author, and as he walked out someone asked whether we had any homework that week.

Not missing a beat he said, "No, there is no assignment this week, as I assume that, being the products of the American school system, you will all have read *The Education of Henry Adams* several times."

After class I went straight to the nearby Strand bookstore and asked one of the clerks to help me find a used copy among their endless stacks of books.

"Any idea who wrote it?" I asked him.

"I'm not sure, but I think we'll find it in fiction."

After some futile searches we finally found it in the Americana section.

"Turns out Henry Adams wrote it," he said, handing a lightly worn copy to me. We both laughed, acknowledging our shared ignorance.

It was one of those moments that sticks with you. I went home and wrote Hitchens a letter questioning my liberal private school education, which had skipped so many of the classics we were reading in his class. I handed it to him at the next meeting, saying, "I just decided to write this."

I hadn't thought of it as an assignment or an essay, but the following week he handed it back to me with an A− written on the top. That was the beginning of what became an important relationship in my life.

I loved Hitchens for his humor, intellect, and kindness. His very complete humanity, really. Being in his presence I always learned something new.

I read a lot about Cuba during those days and also Iran, where my dad had just begun returning once a year after not being able to visit his three sisters who still lived there since the revolution in 1979. It seemed as though Iran was turning a corner as they had just elected a reform-minded president, Mohammad Khatami, who signaled a desire to open up to the world, including the U.S.

As I went deeper I was more struck than ever by how one-dimensional and weak the news coverage of these two places, which held very different places in my heart and mind, was. Both sucked for the same reason.

I was mulling that over one day in 1999 as I read the foreign section of the *New York Times*. On the same page there were stories about the Castro regime in Cuba, and Khatami and Iran. I looked at them both. They weren't long, but the language was similar. In both cases the writer painted these countries in much harsher tones than they would have used for nations with less adversarial relations with the U.S.

It gave me an idea, which I took to Hitchens.

All seniors were required to produce a thesis-length research paper to graduate. I asked him if he thought there was something of value in comparing U.S. news coverage of Iran and Cuba before and after their respective revolutions. He called it a "latent and absorbing polemic," and I got to work.

We talked about our experiences in Cuba, he since the 1960s and me much more recently, and also our shared desire to visit Iran. He had been denied entry by both the shah's and ayatollahs' regimes. When I was finally able to go in 2001 Hitchens encouraged me to do whatever it took to report from there, and he continued to open as many doors as he could for me in the years that followed.

Just before moving to New York I had told my dad, vaguely, that I "wanted to write."

"What do you think you're going to write, and who the hell do you think will read it?" was his chortled response.

I finally had my answer. Well, to the first part, at least.

If it wasn't for Hitchens, none of this would have ever happened, I told myself every once in a while, and not in an angry way.

I began to feel a little less alone. By the end of October my wife and I were allowed a short visit each week. As the weeks dragged on we discussed the need for more public exposure of my detention. Through its actions, Iran's system makes clear that the value of a human life is very little; the only thing less valuable is a person's time.

We decided that trying to get as much attention as possible on our case was the best course of action. I was effectively silenced in the prison, so all the work would have to be done outside.

I told her, soberly, that if we chose that path I would certainly be freed one day, but that it might mean that it would take much longer.

But at that moment we made the decision that we wouldn't be able to live with a headline that read, "Rezaian Pleads Guilty; Sentence Commuted."

8

Let's Face It: I'm a Hostage

I had been in Evin for over ninety days when Yadoallah arrived. Mirsani and I thought we had been
there for a long time. When Yadoallah informed us he'd
already been in section 2A of Evin Prison for over two
and a half years our hopes dropped and our worst fears
were confirmed. We had been moved from solitary because we wouldn't be leaving any time soon. They put
us into a section of the prison for long-term residents
who, for reasons never explained, could not mix with
the general population. "It's not safe for you there,"
was the stock bullshit answer from the prison staff.

Yadoallah didn't like the room we were in. He saw
it as a step backward, because he had also lived in it for
the first year after he came out of eight months in solitary. Then he had been moved downstairs to a larger

room with a bigger and sunnier yard. It was all he talked about, until finally the warden agreed to move all three of us to that space.

On the morning of November 4 something strange happened. It's a date that I'll never forget. How could I?

Due to a difference in the number of days between the Islamic lunar calendar and the accurate one that the rest of the world—the world that has long understood and accepted that the Earth revolves around the sun—uses, November 4, 2014, happened to be both the anniversary of the 1979 invasion of the U.S. embassy— the Islamic Republic's original hostage-taking act, and the basis for their trademark brand of crimes against humanity moving forward—and Ashura, the most important Shia holiday, the commemoration of the martyrdom of Imam Hussein, the sect's central hero.

Imam Hussein, a descendant of Muhammad, along with a band of several dozen fighters was slaughtered by a better-armed force of a thousand followers of the caliph Yazid. The day marks the beginning of the simple yet irreversible divide between Sunnis and Shias that most non-Muslims refuse to understand.

This, to Shias, is the moment that defines their worldview, the idea from which Iran's theocracy draws whatever appeal it still possesses. It's "us against the world" in the classic sense, and fourteen hundred years later all Iranians need to do is turn on the local news to hear that message repeated for them. And on that day more than any other.

The television was on when I woke up. The state broadcaster had two of its favorite annual flash points to cover on the same day. *Will it be flag burning and death to America, or thousands of black-shrouded, chest-beating wailers in unison?* No need to choose, we can have both at once. They're both intended for the same audience.

I watched, disgusted but enthralled. Understanding this spectacle and others was always part of the reason I had come to Iran. I would have done anything to cover that day's events. I'm sure I could have found an angle that would have turned the whole thing on its ear. In a way, it's what I loved about the place. *Better not to let myself go down that path so early in the day.*

Before long a guard arrived.

"Sixty-two, let's go," he said, preoccupied with the mourning rituals like everyone else. We could hear the chants and prayers being repeated in the prison mosque, just on the other side of our tall wall. It was a day when most Iranians, Shia ones at least, put aside their differences.

The guard led me a few feet from our cell's door into the open. Kazem was there.

"Salaam," he said.

"Salaam. My condolences," I said, knowing that it was the right greeting for the day. That he was dressed in all black, with several days of facial-hair growth, was a good reminder.

"And to you. You are Shia," he pointed out.

"Well, I'm not Joe-ish," I said, and he laughed. "Interrogations even on Ashura? *That* can't be legal," I said, surprised.

"No, no. The Great Judge thinks it will be good if you exercise. Twenty minutes every two days."

I wasn't sure what to make of this offer.

"It is a big sign that you are going home," Kazem said. "You must be healthy."

All things considered, that made perfect sense. After he'd failed to meet all the promised release cutoffs, here he was, ashamed, telling me that his superiors said I could work up a sweat. I wasn't about to argue.

The gym was a small room, obviously underground. There was a bench press, some weights, and a rickety-looking stationary bike, cross trainer, and treadmill. The room was lit by two exposed bulbs.

I pondered my options. I hadn't spent time in a gym in years, and I was feeling extremely weak. *Twenty minutes isn't much,* I thought as I mounted the cross trainer. It felt good to move my flabby muscles. I tired quickly, and I sweated. Way sooner than I wanted, the time was up.

The guard had a camcorder and recorded my exercising, documentation that I was being treated extremely well.

"We'll bring you back the day after tomorrow," the guard told me, leading me home.

I had turned some kind of corner. I watched old footage of the embassy takeover and the flag burning, and knew all the lies embedded in the official state narrative as well as the people responsible for taking the hostages. The symbolism of the day and the minor gesture mattered, I just didn't know why yet. I was now an undisputed point on a long line. For the growing community of us—those foreign nationals taken captive by the regime—Iran's revolution was actually the birth of the Hostage-Taking Republic of Iran.

The Iranian revolution culminated in 1979 with the founding of the Islamic Republic and the taking of the American embassy in Tehran, that 444-day episode that consumed the nation's attention. Nearly four decades later Americans still know what "hostage crisis" is shorthand for, and to the small but growing diaspora community of Iranians living in the U.S. it was certainly a crisis.

I can't say I remember that period well. I was three

when it started and not even five when it ended, but it left a deep and indelible mark on our lives.

My dad had already lived in America for over twenty years. He was a citizen and a well-known member of the Marin County community. He was anything but fanatical. He embraced almost everyone he came in contact with, sometimes physically.

He had the thickest accent, and when people would invariably ask him where he was from his response was, "I'm Iranian by birth and I'm American by choice, and I'm proud of both." Such a simple, honest, and maybe uniquely American attitude to have. One that some people—including very powerful ones in both his native land and his adopted one—have been trying to stamp out, unsuccessfully so far, since that time.

Life became more challenging. He used to tell me that in the month before the hostages were taken, October 1979, he did $400,000 in sales. And that's in 1979 dollars. But once they were taken he didn't sell a single item for the next six months.

Iran had been one of the U.S.'s strongest postwar allies. Just two years earlier President Jimmy Carter had visited Tehran and toasted Iran and the "great people of that country who are our close, present, and future friends."

The atmosphere for my dad and our extended Iranian family altered very suddenly and permanently. People he had called friends for a decade or more stopped talking to him, his livelihood was effectively eliminated, and he and other relatives were harassed, were threatened, and had their property vandalized.

One morning when he arrived at his shop he saw a hole in the front window; inside he found a bullet that had gone right through the glass. He never replaced that pane or plugged the hole. "I keep it as a reminder of what people are capable of. Even here," he said.

In the middle of the hostage crisis and all the drama it caused in our lives Dad's lawyer called with some "good news," which my dad was eager to hear.

"Taghi, one of my other clients heard through the grapevine about all the trouble you're facing and they want you to know they've got your back."

"Who's that?" Dad asked.

"The local chapter of the Hells Angels."

He always laughed when he told that story.

We lived in what was then and still is considered one of the most progressive communities in this country, but for him it became antagonistic.

But Dad was undeterred. He had a boundless optimism that wouldn't quit. "God is great," he would say loudly whenever things seemed to be at their worst. But he struggled to keep his business, and an extended family that relied on him for so much, afloat.

As the revolution took hold in Iran more and more relatives made their way to California, many with no more advance warning than a phone call from SFO that they had arrived, some of them living with us for months at a time. The door remained open and the fridge stocked. It wasn't sustainable, but it was what he knew, and my mom never flinched in her role in helping new arrivals acclimate to a life in America.

On January 20, 1981, the 52 American hostages held in Tehran were finally released. It had been 444 days. As all Americans did, my dad felt genuine relief.

But his was different. Hundreds of American business owners offered gifts to the returning hostages. There were free steaks for a year from Nebraska, trips to Hawaii from the Aloha State's tourism board. Major League Baseball presented each returnee with a golden ticket that provided entrance for the hostages (and a guest) to every regular-season baseball game for the rest of their lives.

And my dad offered each one of them a rug. In fact a $1,000 gift certificate for rugs. More than forty of them took him up on his offer. Some of them came in person, others sent local friends, and some just phoned in orders and my dad mailed them a rug based on their desired parameters. Along with each one of these rugs he put a certificate in the package that said, "As an Iranian I apologize for what you have endured and as an American I welcome you home."

News of that gesture spread. A wire story about it was picked up in newspapers all over the world. Some weeks later he received a postcard from Kathmandu from a hippie who used to hang around his shop. The news had reached that far.

Charles Osgood included him in a segment of his popular rhyming current events feature, "The Osgood Files," at the end of CBS's Sunday morning news broadcast; they spelled his name wrong, but that was my dad.

And of course Iran's revolutionary government took notice, too.

For years every time a Rezaian would return home to Iran there were questions at the airport about Taghi and what he was doing. That was natural, because at that point very few Iranians in America had any public profile.

I thought about that episode as I moved my bones on the elliptical machine and tried to hold on to whatever dignity I still had.

The gym, it turned out, was my refuge, which was

unlikely as I hadn't done any honest exercising since college.

It was in there that guards would ask me their own personal questions and I, if I was feeling up to it, would hold court. Always one at a time, because they were too afraid to speak openly in front of one another. I soon learned that it was the only place indoors I was taken to that wasn't wired. Still, I watched what I said.

We were in the new cell for less than a week when Mirsani got word that he was leaving. He thought he was being freed, but it turned out he was only getting sent to the public ward.

So now I was back to a life with one other person.

Yadoallah had presence and we became fast friends. He was loud and he sang. He joked with everyone, was mostly polite, prayed five times a day, and was one of the few people I knew who could eat more than me. I loved that about him. By the time he was brought into the cell I was down forty pounds, and he weighed about two hundred fifty.

He chain-smoked, and told me he had never had a cigarette until he was forty.

He was a near-beer mogul whose only crime was being born Sunni and Kurdish; both were seen as threats to Iran's ruling Shia establishment. He stayed in prison because he refused to cooperate with the IRGC;

they were trying to shake him down. "How can I lie before God about things I haven't done? I won't sell the next world to anyone," he told me.

He was optimistic, but the isolation was making him crazy. He had four kids, who were living a luxurious life that he never could have imagined for himself growing up in war-ravaged Kurdistan, a western region on the border with Iraq that was bombarded throughout the mullahs' eight-year war with Saddam Hussein.

The weather began to cool and soon I was allowed an hour in the gym every day. Yadoallah was a good companion, because as bleak as things were, he always looked on the bright side.

But on Thanksgiving Day, I woke feeling especially lonely. It was the first real holiday during my time in detention.

Kazem arrived at the cell gate and said, "Salaam, J. Let's go. We haven't much time," which is something he said often. In the beginning he said it so much, it was the first indicator that he wasn't being honest. If we had plenty of anything in there it was time, and he and his cohort, I learned, were people willing to wait.

Kazem hadn't arrived alone. Rezvan, a guy who, while I was still in solitary, had interrogated me about the many pictures of my life found on my computer, was there, too. He wore a Kangol hat and sunglasses over a surgical mask to hide his appearance, and he was taller than Kazem. He hadn't been to see me in over two months. "Was your son born yet?" I asked, remembering that his wife had been very pregnant, or so he said, several months earlier.

"Yes. Good memory."

"Well, it's part of my training."

They cracked up, knowing I was nobody.

They led me to the interrogation building, which was quiet. It was a late morning.

Kazem produced a Ziploc bag with Yegi's and my cell phones in it, including older ones that we hadn't used in years. Each of them had a tiny white sticker with a number in the Arabic script used in Iran written on it. To Kazem and his team they were evidence, but I felt like I was looking at artifacts from another era. They seemed almost make-believe.

"Which one is yours?" Kazem asked.

I pointed to the black iPhone 5 that the *Washington Post* had bought for me.

"Turn it on," he commanded.

I wondered what was happening.

"Call your mother," he said.

"What? Why?"

"You must call your mother."

"And tell her what?" Yegi was the only person from the free world I had any kind of contact with and I was startled by even the idea of expanding that circle.

"Today is a holiday and you must congratulate her. She is your mother. It is the Great Judge's gift to her."

"No." I don't know why I resisted, but I did. "I'm not calling."

It could have been that I didn't want to cry on the phone to my mom, whose voice I hadn't heard in months. Or it could have been that I wanted to tell the Great Judge to take his gift and shove it up his ass. Whatever the reason, I said I wouldn't do it.

"You don't have a choice," Kazem said.

This went on for a few minutes and then he said, "Your mother wants to hear your voice." He paused. "She's your mother."

With that, I relented.

My own phone didn't work. The bill hadn't been paid since my arrest and service was cut.

Rezvan handed me a prepaid calling card and his very basic and old Nokia. These fossils were popular in Iran and other Asian capitals—retro digital. I punched in the card number, password, and my mom's number. Nothing happened. And again. Nothing. Finally it rang, but there was no answer. This went on for several attempts. They were getting antsy.

"Where is your mother?"

"How would I know? You haven't let me speak to anyone in four months."

"Keep calling. She must answer."

I tried another number. It was her cell phone and it rang momentarily, then I got a message in Turkish, probably saying she was out of range. And again.

Finally it rang and was answered. It was my mom on the other end of the line.

"Hi, Mom," I said, trying to sound as calm as possible, fighting back a rush of tears and the biggest lump I've ever had in my throat.

She was equally cool and collected, and I appreciated that.

"Where are you?" I asked.

"I'm at Cousin Barb's," she said. Her cousin lives just outside of Chicago and has a massive Thanksgiving gathering at her house every year. We used to go when we lived in Illinois. It was the one place I wanted to be right then, surrounded by people who'd known me since childhood. I thought about how weird this must all have seemed to them. They knew my dad and his sisters. Their own extended family had experienced plenty of tough times. But this was different. Other-level stuff.

"You want to say hi to anyone?" my mom asked.

"No, Mom," I told her. "I just called to talk to you."

We talked and talked, but with Kazem and Rezvan hovering over me, I knew I couldn't say all that I wanted. They couldn't understand our conversation completely, but I spoke fast and colloquially just in case. And made it obvious to my mom that she wasn't getting the full picture. I kept referring to Kazem, like, "My interrogator says I should tell my mom that they are treating me very well."

"I'm sure he does," Mom would reply.

And: "My interrogator says I am very sorry for all the suffering I have caused you."

"Put him on the phone right now! You didn't do anything and he knows it. I want to talk to him," she said.

I looked at Kazem and said, "She wants to talk to you."

He was embarrassed. He didn't do well with women and talking to my mom was definitely not part of his "mizhan" for the day.

"I cannot."

"He says he can't." I was playing telephone, on the telephone. Then to Kazem, "My mom wants to know when this will end."

"God willing tell her it will end very soon," he said.

"She wants to know when." She hadn't asked that.

"Very soon."

"Sure?" I asked. She wants you to swear to God." She didn't ask for that, either; I did.

"I swear to God," Kazem said.

"Mom, did you hear that, he just swore to God that I'd be released very soon."

"Good. That's a start." She and I both laughed.

We tried to keep it loose and I asked about everything I could remember to. The call went on. At one point I looked at the phone and it had been well over half an hour since it started. I couldn't believe it. Maybe there was some reason for hope.

"Mom, is anything actually being done to get me out of here?"

"Oh yes, honey. A lot." But she couldn't tell me what.

It was time to wrap. I didn't want to end the call that I hadn't wanted to make in the first place. This is how it goes sometimes. I was worried I might never hear or see her again. Really, how could I know?

"You should come here, Mom," I pleaded with her. "You really should."

"I know."

She could handle it, I was certain. She had seen plenty by then.

My mom grew up in Warrenville, Illinois. She was the only child of Drs. John and Mary Breme, an unlikely pair who had divorced by my mom's fourth birthday. Dr. John was born in Italy and came with his family as a small boy to search for a better life. He became a doctor and taught biology in Chicago. One of his students was my grandmother, nearly thirty years his junior. Sensing her potential, he put her through medical school.

They were not well matched.

Dr. Mary, whom we called Nana, left Dr. John and took my mom, little Mary Ellen, with her.

It was the mid-1940s and Catholic single mothers in and around evangelical Wheaton, Illinois—Billy Graham country—were a rarity. Rarer still was the fact that she was the town doc. She built a clinic there to which people would come from long distances because of the reportedly miraculous results her patients experienced. To this day I meet people who know that she was the doctor presiding over their birth. She was a big deal.

She never remarried after her divorce, and raised my mom with the help of her older sister Mae and Mae's husband, Cal—"Unk" to my brother and me—who were unable to have kids of their own.

Once a year Nana would take my mom on a great adventure. There was the dude ranch in Arizona where they were regulars, an epic and calamitous journey by horse into the heart of Glacier National Park, and a trip back to Slovenia to visit their remaining family there. If Dr. Mary couldn't take the time off it didn't stop her from sending my mom to see the world, like Mom's Atlantic crossing on the *Queen Mary*.

Back home my mom raised farm animals, and she was an early adopter of Spanish.

After graduating from Wheaton Community High School—a year ahead of my *Washington Post* colleague and journalistic hero Bob Woodward, and several years before one of my comedic heroes, John Belushi—my mom enrolled at Westmont College, a very conservative school that made students sign a contract promising to never dance with boys or drink alcohol, but happened to be in freewheeling Santa Barbara, California.

It wasn't long before she realized she had grown beyond the narrow scope of strictly Christian environments, like the one she was raised in.

First she signed up for a dig—she was an archaeology major then—in what people used to call the Holy Land. It was Palestine back then, or maybe Jordan, but someplace that is now part of Israel. She was a white chick infatuated with the Middle East decades before that was even a thing.

When she returned to the U.S. she made a calculated move to start fresh, transferring to San Francisco State University just in time to fully experience the sixties.

One late afternoon in the library, my dad, the young man from Mashhad, Iran, saw this adventurous girl from Illinois.

After some glances between them he finally went to her and said, "Coffee."

"Are you asking me or telling me?" was my mom's response as she tells the story half a century later.

Their relationship may sound unique for the time, but all around them were other Iranian guys dating American girls—not only were they enthralled by these exotic young women they found, there also weren't many Iranian girls studying abroad yet.

Mom even decided to learn Farsi, trading language lessons with other new arrivals from Iran. Very quickly Iranians became the largest population of foreign students in the U.S. and many of them came to San Francisco.

When my mom told Dr. Mary that she wanted to marry my dad, my grandmother said she wouldn't give her blessing. Nana suggested they visit Dad's family, assuming my Mom would be put off by their foreignness; Mom saw it as an opportunity. Soon enough these two ladies from Illinois were on their way to Iran to meet my dad's family. In 1967. That should tell you much about the two of them, but also about how the world has changed since then.

I still have pictures from that trip of my mom and two of my aunts, who years later would move to America but were teenagers then, in miniskirts and beehive hairdos.

When they arrived in Mashhad my grandmother asked her most pressing question of Hajj Kazem, which was designed to be a deal breaker.

"If my daughter marries your son, will she have to convert to Islam?"

She wasn't prepared for my grandfather's response.

"As Christians, you are also people of the book," he told her, using the Arabic term that refers to the single God shared by Judaism, Christianity, and Islam. "So that is not a problem. But more important, it's the two of them that will have to lay down together each night. The decision is theirs to make, not ours, no?"

My grandmother was impressed, but remained skeptical.

Now, I'm not sure if my grandfather treated his six daughters and their various suitors in the same way, but such a forward-thinking attitude at that time was a remarkable thing, and I believe it influenced how my parents raised my brother and me years later in Marin County, California, about as far away from Wheaton and Mashhad in terms of a worldview as one could imagine.

Neither of my parents responded well to authority or conventional wisdom. They both had goals they wanted to achieve and they didn't appreciate anyone standing in their way.

Soon after they were married my dad was working in Emporium Capwell, a popular San Francisco department store, where he managed two of the departments, toys and women's shoes.

Around that time my grandfather had sent a small shipment of Persian rugs to San Francisco. This had been one of the family businesses back in Iran. Needing more money to support their life, my dad placed a classified ad in the local paper and was pleasantly surprised at the speed with which all of the rugs were sold. He asked my grandfather to "send more rugs!" Soon a shipment many times larger than that first one arrived, and my dad recognized an opportunity.

They moved right across the Golden Gate Bridge to Mill Valley, at the base of forever-green Mount Tamalpais, where my dad set up a Persian rug shop, one of the first in that part of the country. And it was an instant success.

In typical immigrant fashion, he was holding down two jobs: one for the Man, and the other for himself.

One day, tired of the commute, and confident of his business's growing success, he asked the Emporium for a modest raise, to which he was entitled given his experience in the company. His boss refused to give him one.

He quit on the spot.

That night he went to my mom and told her, "Hon"—that's what he called her—"our life may not be easy and we may have to eat canned beans for a while, but I promise it will never be boring."

"That's why I married you," she replied, and in my mind that's when their life—actually, our lives—really started taking shape.

It was 1969 and hippies, especially the ones with money, of whom there were many in that area, loved rugs. And by extension they loved my parents. At that time there were no biases toward Iranians or Muslims, because there were still too few of them to judge.

If anything, my dad, his fellow countrymen, and their culture were admired and appreciated for being exotic, ancient in their traditions while worldly in their ingenuity, and hospitable. If one distills the many stereotypes, that's how Iranians have been known for centuries. I always tried to recapture that unique national character in my reporting.

My mom knew Iranians and she knew Iran. She wasn't afraid, but I realized she might have been underestimating what bastards I was dealing with.

We said our goodbyes, and I hoped she'd hightail it to Tehran.

After we hung up, the phone at Cousin Barb's house rang again. It was Secretary of State John Kerry.

He'd told Iran's foreign minister, Javad Zarif, in their last meeting that a Thanksgiving call would be well received. Apparently Zarif had enough pull to arrange it. Secretary Kerry had also written me a letter to lift my spirits, but that was never delivered.

The State Department—via my brother, via Yegi— knew my morale was low, but they wanted to send me a message that I hadn't been forgotten. Some, but not all, of that got lost in the game of geopolitical telephone.

Around that time the State Department and Iran's foreign ministry had agreed to start a secret set of negotiations for a prisoner swap. Obama's envoy on ISIS, Brett McGurk, was chosen to lead the American delegations. That phone call was the first act of good faith.

I had been locked up longer than any of the previous

dual-national journalists or scholars, and there started to be little signals that I might be released soon.

In an early December interview with France 24, Mohammad Javad Larijani, the Berkeley-educated head of Iran's human rights commission and older brother to the speaker of parliament and the head of the judiciary, called my detention a "fiasco" and predicted I'd be out within a month.

Christmas 2014 fell on a Thursday and it had been two weeks since all of my contact with the outside world had been cut for getting caught trying to smuggle a note out of prison with Yegi during one of our visits. I had successfully gotten three out so far, and got cocky. The guard on duty that day was apparently paying attention to the surveillance camera for once.

Kazem's boss, Borzou, told me that getting busted for what appeared to be such a silly mistake was an act so clever that they still couldn't figure out why I did it. Maybe they were right.

I'd heard rumblings, through Yadoallah, who'd heard from his family, that the international calls for my release were growing louder. Some of the guards confirmed this. If that was indeed the case, the silence from me may have been compelling my family and employers to take more public action. But that was just a guess.

Around eleven A.M., Khosrow, a little guard with a nervous tic—he was always shooing away an insect that wasn't there—came to the gate of our cell and told me to come with him. It could have been anything: interrogation, freedom—it *was* Christmas—or Yegi. Whatever it was I didn't like the not knowing.

He led me in the direction of the visitors' meeting room so I guessed it was my wife. When I entered the room I paused with mild shock. It was my mom and mother-in-law. No wife. I was happy to see my mom after so many months, but not ecstatic.

It was a hard meeting. My mom was trying to be upbeat and my mother-in-law was nervous but also frustrated; this was not a life challenge she and her family deserved. And I was just angry. More than anything because I was helpless but feeling responsible.

They gave us twenty minutes. I implored my mom to do more to get me out. "These people are crazy," I told her.

"We're doing everything we can, sweetie," she assured me.

"Do more. Faster. I'm going crazy in here."

When Khosrow came to take me back to my cell I was even angrier. "Why didn't you tell me it was my mother?"

I would have prepared myself more, mentally. I could have explained that to him, but he wouldn't have understood. And I definitely would have brushed my teeth.

"It's Christmas," he said, with a hand on my shoulder. "I thought it was a good surprise."

Yadoallah had already been taken to his Thursday meeting with his family by the time I returned to the cell. At a minimum I had a Kurdish feast to look forward to.

His wife would go all out on the days of visits as if it was a celebration, creating the most elaborate meals imaginable: eggplant stew with lamb shank, stuffed grape leaves with lamb shank, meat and potato patties called *kotlet.*

But today was a special occasion. His family had gone to Dubai to take care of some business there, and they had brought back food from his eight-year-old son's favorite fine dining establishment.

I saw the golden arches on the bag immediately. I already knew that in that part of the world McDonald's represents an idea that it could never live up to in real life. It's freedom, speed, and shopping malls. In Iran, where American brands and chains have been officially banned for decades, it takes on extra significance. It is the one restaurant name that every Iranian—whether they have ever tasted its forbidden fruits or not—knows very well.

No negative connotations whatsoever. It's a lifestyle thing. If you travel anywhere in the world it's the first restaurant you go to. If there's not a McDonald's there, it's not a place worth visiting. It's not something you hide from your friends and family. This is no guilty pleasure. You take pictures of your outing to Mc-Donald's and you post them on Instagram.

McDonald's is so seductive that there is a whole body of Iranian sociological work that rails against the *Mc-Donaldsization* of societies. It is, for none of the same reasons we consider it to be here, the greatest evil. The culinary embodiment of the Great Satan.

When Yadoallah reentered our cell and placed the contents of two McDonald's bags on our small plastic table, six Big Macs and four cans of Coke, I knew I was about to give the biggest symbolic middle finger to the Islamic Republic I possibly could in my current predicament. And it was Christmas *fucking* Day.

I put down two Big Macs and a Coke, never even considering that it was anything but the most righteous act I could perform just then.

9
2011

But nothing happened after that Christmas visit. I was in a holding pattern: no court, no visits, no light at the end of the tunnel. This was a jam, no doubt, but I tried to remind myself we'd survived worse.

Two thousand eleven was my hardest year. Since I'd returned to Tehran in the fall of 2009 my freelancing career had taken off. I was attracting new gigs and cementing my place in the foreign press corps in Iran.

I wrote Christopher Hitchens a letter. His cancer had progressed dramatically at that point, and it was important for me to tell him something he already knew: that he had had a profound impact on my life. I told him, also, about my relationship with Yegi, and how being so close to a young Iranian woman made me simultaneously irate over the treatment of females, but also enthusiastic about Iran's prospects, as it was inevitable that one day women would run the show.

The Persian New Year was coming and I made plans for Yegi and me to go to Europe. It was all a part of an elaborate scheme to get her into the U.S. on a tourist visa, which we had failed to achieve a year earlier. I knew that if she was given, and then used, a Schengen visa, which allows entry into the eurozone, and returned to Iran before the end of the visa's validity, her chances of getting an American one would shoot up exponentially.

I had become friendly with the number two man at the Slovenian embassy, who considered me a fellow countryman—I'm a quarter Slovenian, on my mother's side—and I was exploring opportunities to send young Iranians there. Slovenia was one of the few EU countries granting Iranians visas at the time. Yegi was a perfect test case.

Our relationship was advancing and it was important to me that she start seeing more of the world. Ultimately I wanted to take her to California, first to get to know my family and second to see if she would like it there as much as she expected she would. I had seen the disappointment before of young educated Iranians coming to the U.S., bored by how slow and simple most of it can seem compared to frenetic Tehran.

Meanwhile, I was still enjoying the opportunities of an expat American in a part of the world where we are coveted.

In late January an old high school friend of mine who was working as a lifestyle reporter for a media company in Dubai asked me if I'd fill in for him on a gig because his wife, another high school friend, was days away from giving birth to their first child. *Why not?*

I was flown, business class, to Cape Town to test-drive the new BMW 6 Series convertible, and put up in a six-star resort at the base of Table Mountain. This is the sort of one-off thing that happens when you can complete a sentence on deadline and live in the Middle East.

I wasn't even considering going back to a life in the U.S.

On the evening of March 4 I got an email from my brother asking me to call him. I was headed out and wrote back to him that I would be gone for a couple of hours and would like to call him the next day if it was all right.

"I need you to call me tonight," he responded.

My dad was seventy-one and had been struggling with health issues compounded by a car accident he had in 2009, a couple of months before my move. He'd had a quadruple-bypass operation when he was forty-nine and began to take his health very seriously then, including an extensive exercise regimen, which was disrupted by the car accident.

I braced for what I thought was the worst.

I called.

"Hey, Ali, what's going on?"

"Hi, Jason. I'm at the hospital. Walker wasn't feeling well this morning and—" He didn't continue and I butted in.

"But he's okay now, right?"

"They brought him here . . ." He trailed off.

"Okay. So what's going on now? He's all right now?"

"He died."

No matter what else I tell you, know that this was the single hardest moment of my life.

Our call didn't last much longer. My not-even-six-year-old nephew, the sweetest, brightest little man, was gone. I don't break down emotionally, but I wailed and beat the wall with my open palm and then fist. It was the first time in many months that I'd felt far away.

I called my mom.

"Mom, what's happening?"

"Oh, sweetie. I don't know what to say." She was composed. She keeps it together better than anyone I know, including me.

"I just keep thinking about what you would have

done if this happened to Ali or me," I said through my sobs.

"Well, you need to get those thoughts out of your head. That's not what happened, and we need to be there for Ali," she said in a tone I don't remember hearing very often. Kind but forceful.

"You're right," I sobbed. "I'm coming home."

I departed Tehran out on the next possible flight and arrived back in Marin County the following night. It was horrible. Walker had been hit by a fast-moving H1N1 virus that was too much for his immune system to combat. His death was shockingly sudden.

What is there to say? We were, as a family, completely shattered.

After thirty-eight years at the house on Oak Ridge Road my parents had just moved into a small condo. My dad had run out of luck and couldn't sustain the massive financial burden of keeping the place afloat, having leveraged the mortgage once too often for cash infusions into his rug business. The new owners hadn't taken possession of it yet; the only piece of furniture left was a bed in my high school bedroom. I decided I wanted to stay there until the last possible moment.

This place that had meant so much to all of us, I realized, was just a shell. All the energy that came from its being lived in was gone. It was freezing cold and the late winter rain pummeled down. The cascading sound large drops made on the living room's high and arching roof had been one of my favorite childhood sensations. I cried and cried.

It was becoming clear that Walker's death, along with losing his hacienda, was the beginning of the end for my dad. While the rest of us tried to handle the situation with whatever grace we could muster, he was being carried away into heartbreak. A couple of days later he had a heart attack.

When I got to his hospital room and sat down next to his bed, he told me that one of the nurses had been there when Walker was brought in and had described the grief that the entire hospital shared that day. I wanted to kill that asshole for adding to my dad's sorrow.

I didn't tell Ali at first about the heart attack, but then realized it was worse to hide it.

We had spent too much time at Marin General Hospital over the years. Hajj Kazem had died there and Nana had spent so many stretches there that we'd lost count. We hated that place.

I talked to Ali about Yegi and told him about our Europe plan.

"You should go, and then come back if you can. I'm probably going to need your help here," he said.

"You sure?" I asked.

"Yeah. Go."

We didn't really know it then, but it's the weeks and months that follow impossible-to-fathom losses that are the most gut-wrenching.

I was in a prolonged haze. I made the decision that I had to keep it together, whatever that looked like. Obviously I couldn't sleep.

The rain didn't stop much those few days.

Ali was doing a contract job in San Jose at a friend's business. He decided to go to work in the middle of the week after Walker died. I drove with him down there one day. I couldn't believe he was even able to get out of bed.

"It's good that they have another child to raise," a lot of people said.

After his heart attack my dad was only in the hospital for a couple of days. "If he stays there any longer he's gonna die. It's too depressing," I told my mom. They released him.

I spent as much time with him as I could, because I thought I could bring him back. I had seen his long and dramatic mood swings before, but it had been years since I had seen him low for such a prolonged stretch. He wasn't saying "God is great" much.

His neck was in a brace due to the car-crash injuries and his mind was somewhere else.

One afternoon he asked me to drive him to Petaluma, a town about twenty miles north of where we lived, where he had had a shop since the mid-1990s. That was where I had done most of my training in rugs.

He still had the "rug annex," as we called it, a funky wooden building that used to house a local radio station, which was next door to the palatial old bank that the store had inhabited for the past fifteen years. The annex was the opposite of the former bank, which he called "Monarch." More of a hideout in the shadows of his last shop than an actual business. I hated it there.

We walked in and I was overwhelmed by the number of rugs. By then I understood this business as well as anyone could and realized that some of the inventory had to be there on consignment. I could easily recognize styles that my dad wouldn't invest in, but would stock, and other trends in items he might buy at bargain prices in an auction, at an estate sale, or as a trade-in. There was simply too much stuff.

I couldn't stand being there and I took my dad home.

Just like when we first met, twice a day, once in the morning and once late at night, I was speaking to Yegi by phone. She was an absolute rock for me when I most needed it.

"Stay as long as you have to," she told me. "I will be here."

It was one of those periods in life that either cements or unglues a relationship. I felt lucky.

She tried to convince me that it wasn't the right time for us to go to Europe.

"No, we have to go," I told her. "I'm not sure how long I will be stuck here and if we don't use this visa who knows when we'll be able to get you another one." I don't think of myself as a very stubborn person, but some decisions stick. It helped that Ali and my mom encouraged me to go. Dad was still too distraught to make decisions or even have an opinion. I was just going to be gone for nine days.

My dad sat on a chair in the living room of this unfamiliar place—there was wall-to-wall carpeting and not a single Persian rug on the rented floor—watching the devastating news of the Fukushima tsunami. Maybe it took his mind off of his own misery. I don't know.

I kissed him goodbye on both cheeks, not hugging him too hard.

"We need you to get stronger, Dad. Take it easy the next few days."

He just nodded.

As I walked out the garage he said, "It's good that you came."

"I'll be right back," I said, smiling, trying my best to lighten the mood.

It still hadn't totally sunk in that life as we knew it, the one that my parents had begun cultivating almost half a century earlier, was disintegrating.

I flew that day from SFO to Milan, where Yegi would be waiting for me. She had flown in on AZAL, the airline of the Republic of Azerbaijan, via Baku. It was the only ticket to Europe from Tehran we could afford, as fares during the Iranian New Year high season, which was right then, routinely triple in price.

I had booked us a room at a place called the Hotel New York, the one place in Milan that fit my third-world budget.

It was late night when I arrived, but Yegi was waiting up for me. Seeing her after the toughest weeks helped me relax. I knew she was my future.

We were two tourists, among the thousands, who weren't exactly on vacation. We walked through the Duomo, Milan's massive cathedral. Yegi's first-ever church. We took a bus to Venice, and then another one to Slovenia, where we visited my remaining relatives there.

Finally we made our way to Paris, to spend time with old friends of mine.

Those days are hazy. I would cry whenever I felt the need to, and Yegi would give me a look, head cocked, that simultaneously said "I'm here for you" and "Don't cry . . . so much."

I took her to the Rodin Museum and we sat in the sculpture garden, always one of my favorite places.

On our last night in Paris before Yegi was to return to Tehran and I would head back to San Francisco, our friends Charley and Marie, whom we were staying with, hosted a dinner at their place with some of their local friends I'd come to know over the years. Everyone knew Ali and my sister-in-law Naomi. Some of them had met Walker. There were people all around the world who shared great affection for my family. That was a very comforting feeling.

I went to bed prepared for more difficult days but fortified by friendship and love.

The next morning, around six A.M., my phone rang. No number showed up, but I answered it groggily.

"Salaam. What happened to Daie?" I recognized the voice. It was my cousin Samaneh in Mashhad, the daughter of one of my dad's younger sisters, asking about my dad using the Farsi word for maternal uncle.

"He's okay, I spoke to him yesterday."

"Where are you?" she pleaded. "What happened to Daie?"

"I'm in Paris. I'm going home tomorrow. He's fine."

She was distraught. "Okay, okay. Goodbye."

She hung up. I looked at my phone and there was an email from my brother. "You need to call me," it said.

No. I already knew where this was going, but I went into the living room and called anyway.

"Hey, Jason," Ali greeted me unceremoniously. "Believe me, I didn't want to do this again." He was so strong.

"I know. What do I do now? Should I go straight to the airport?"

"Get on the same flight you were going to. No need to change it. There's nothing you need to do today, and you don't want to be here right now." I could hear some of my Iranian relatives wailing in the background. I knew Ali hated to be there and I loved him for being able to sit through that, without me, after all he'd just experienced.

"Spend the day with Charley. I'm glad you're with him. He wants to be there for us. Have a good meal and go see the Monets in l'Orangerie and then take a walk through the Tuileries Garden. Eat something good. That's what I would do if I was there," he advised.

I sat in our friends' living room, on the top floor of an apartment building in Paris's Nineteenth Arrondissement. The sun was rising over Paris. Their place had a complete view of the city, with the Eiffel Tower right in the center of it. I just stared out the window, internalizing for the first time what exhaustion from the pain of sustained grief feels like.

Charley came out.

"You're up early today," he greeted me.

"My dad died."

He heaved a loud sigh and paused.

"J, this is too much."

"Right?" I asked, searching for some validation that this was not fair.

We spent that day just as Ali suggested. I had a hard time keeping it together, but I didn't need to. I was with people who loved and knew me.

They took us to Café Marly at the Louvre. Yegi and I walked around the Marais, prolonging the moments before an inevitable goodbye. We had no idea when we'd be back together. The situation back in California had just gone from unbearable to impossible. At the airport we both cried. I had just turned thirty-five years old. This was not what I'd envisioned that age would be. She got on her flight back to Tehran and I waited for mine to San Francisco a couple of hours later.

I'm never there when my people die, I thought.

Siamak picked me up at the airport and hugged me hard. Ali and I were the older brothers he didn't have. His mom, my aunt Tina, was the middle of my dad's three sisters who had moved to America. Siamak was born in San Francisco in October of 1979, just a month before the hostages were taken in Tehran.

My dad was the latest casualty in the changing of the generations.

I was at peace with my dad's death. How could I not be? My brother had just lost a son who was in kindergarten. Dad was seventy-one and had had not only a full life but a rich medical history, too. There was nothing surprising or tragic about his death other than the timing, four weeks after Walker's.

Siamak drove me to the mortuary. In addition to the shroud—which Dad had bought in Mecca when we went there five years before—and the actual burial there were other funeral preparations, including a multistep ritual washing of the corpse that required several traditional ingredients, camphor and lotus powder among them.

Some of my relatives had been on the phone with local Bay Area mosques to inquire about whether or not they washed bodies for burial. The Sunni ones, it was discovered, wouldn't bathe Shias. Others charged a substantial fee, which was totally reasonable because we're talking about a service, but somehow that didn't sit well with me.

"I'll do it myself," I announced. "It's what he would have wanted."

In fact, Dad had performed this rite for several other relatives, and that's why none of us knew about the hurdles of getting a Shia Muslim corpse in America properly prepared for burial. It's one of those things you don't discuss much.

When I made that decision solidarity among others who appreciated my dad for their own reasons came quick.

Siamak was there, as was my late aunt Mimi's lifelong boyfriend, Bahram. My burial buddies. And Reza Rezaian, my cousin and our neighbor since I was three, when he and his parents relocated from Iran to a house a couple of hundred feet from ours. Until just the summer before on hot days we might come home to find Reza in the pool. That was totally normal. My dad had been there three years earlier to help Reza wash his dad's body. Hassan, another of my dad's cousins— from my grandmother's side of the family—also came, with directions on how to do it properly from one of his nephews who was a mullah in Mashhad.

"If anyone complains about how we did this, tell them we had instructions from the shrine," he said, holding up some notes he had written during a phone call with the cleric at the Imam Reza shrine, where our relatives who died in Iran were all buried.

And Ali was there, which put it all in a new kind of perspective for me. This was absolutely not about me.

After lifting the sheet from my dad's corpse I paused

for a moment, assuming I wouldn't be able to do it. But grabbing his limp and heavy leg I had the immediate understanding that this was not my dad. It was my dad's body, but *he* was no longer there.

It probably sounds stupid, but it was the most comforting feeling of my life.

By the end of the process, which took half an hour, we were all laughing, sharing jokes and memories about my dad.

I realized then that all funeral rituals, no matter what the faith or culture, are designed to force the dead's survivors to let go. I learned later that in Iran loved ones are no longer allowed to take part in this process at most mortuaries.

The next day we went to the cemetery for burial. The number of my dead relatives buried at Valley Memorial, in Novato, California, was getting uncomfortably high. A lot of people came that day, but many others who would have just didn't know, because as a family we were already drained.

We had a slow procession from the chapel to Dad's plot a few hundred yards away. He was being buried alongside Aunt Mimi, on top of a little hill, at literally the highest point in the cemetery.

There wasn't any discussion about it ahead of time, but Siamak, Bahram, and I got in the ground, partaking in the ritual burial rites. Sad, but peaceful, just as I'd remembered. Several of my high school friends who were there reported later that a large red-tailed hawk circled above the entire time. Once we had laid him to rest, with his right cheek touching the dirt, head facing toward Mecca, other relatives above lifted us out. Everyone threw red roses and handfuls of dirt over the corpse, as is customary.

It didn't take me long to realize that the preparation—the washing and the burial of my dad—was the most important act I could perform in life.

I was at a strange peace.

I pondered those days now, knowing that if I could survive them, I could get through this. But I was feeling hopeless.

For months, from where I sat it had appeared as though nothing was happening. My case seemed to be in a state of suspension and it was just life in the cell all the time. I had no contact with the outside world. It had been weeks since the Christmas surprise and my loneliness was becoming unbearable. At least I could still work out, walk in circles in our tiny yard, and read. I poured myself into those three activities to fill each day.

One afternoon in the dead of winter, Yadoallah came back from his daily phone call with his family with a massive grin on his face. "I've got good news for you, J," he said, throwing an arm around my shoulders.

"What is it?" I had to know. In Evin good news was rare.

"Obama talked about you in a speech," he said, as if he were announcing my imminent release. It was late January so it was possible that he'd mentioned me in the State of the Union address, but that seemed an unlikely moment for a president negotiating a game-changing deal with my captors to insert me into the national dialogue. Still, I liked the idea of that.

"Where? What did he say?" I demanded.

"I just know he talked about you and that means you're getting out," he told me, doing his best, as he always did, to focus on the positive.

Somewhere inside I knew he was probably right, but I was finding it hard to believe there was any progress being made toward my release. I couldn't see a single sign of it. It wasn't long, though, until I started to catch glimpses of the cumulative efforts around getting me out.

My name was growing into the latest chorus in the antagonistic call and response between Tehran and

Washington that has defined relations between these two countries—my two countries—for most of my life. I was now a living centerpiece in a struggle that I had spent years, through my explanatory writing, seeking to defuse.

Any time someone in the U.S. called for my release there was an equally weighted damnation of me in the Iranian state press.

As those flashes grew I began to steel myself for confrontation. The verdict in the courtroom was already a lost cause. I knew that. But that wasn't the front I needed to worry about.

My battle was for global public opinion, and I had been winning that one since soon after my arrest, as the drama of the epic nuclear deal played out, shining a spotlight on my case, offering a very sober reminder of why a deal that left me and other Americans in prison might not accomplish what it was supposed to in returning Iran to a sense of international normalcy.

A new acceptance of Iran in the American consciousness as anything but a rogue nation never actually happened, and one ingredient of that was the highly visible and preposterous secret trial against me and the way Iran's supposedly urbane and sophisticated officials were unable to provide any legitimate excuse to support my detention but simultaneously continued to justify it, even in visits to the United States.

And there was another spotlight shining in Washington, one that made it possible for the *Washington Post*'s editor, Marty Baron, to keep talking about me.

"It would be a farce if it wasn't such a tragedy," Marty said of my situation at one of the interviews he gave regarding *Spotlight,* a film about one of his reporting teams that would go on to win the Oscar for Best Picture.

The *Post*'s efforts knew no limits. Although some Iran watchers publicly chastised the paper for not doing more, the company's top brass took on extracting me from Iran by any means necessary as its top priority, with public and secret campaigns to bring me home.

The *Post* published dozens of op-eds by Marty and the editorial board, as well as articles with updates about my condition whenever the most minute new detail was available. There was also a living cartoon by my colleague Michael Cavna, whose drawing of me was updated daily with a fresh hash mark to symbolize the number of days I'd been in prison, which corresponded with a flash on the news ticker outside of the *Post*'s longtime headquarters on Fifteenth Street NW.

Of course I couldn't see any of this, but I started to hear about some of it. Unbelievably and quite absurdly, I was able to see some of the discussions about me on Iranian television, and all the while I was keeping a running record in my mind of all the lies Iranian officials, especially Foreign Minister Mohammad Javad Zarif, told about me and my condition. Or perhaps it's more accurate to say all of the implications he made that could be deciphered as references to me. I was growing to hate him deeply but also respect his ability to bullshit literally everyone he encountered, including himself.

There were many other officials weighing in, as well, some of them calling for my execution to pay for crimes so heinous they could not even be explained, while others in and around the Iranian political system gently, timidly questioned the wisdom of holding a well-known, accredited journalist, very publicly denying him due process and boldly breaking multiple Iranian laws in the process. Doing so, they argued, did little to further the Islamic Republic's interests.

The *Washington Post* and other news organizations kept track of these developments and sought to highlight the protections provided to defendants in the Islamic Republic's constitution that I was being denied. At times it seemed that international media outlets had a better understanding of Iranian law than the Islamic

Republic's judiciary. It seemed that way because it was true.

One afternoon I was watching Press TV and the regular coverage cut to a live one-on-one between Iran's foreign minister, Javad Zarif, and my *Washington Post* colleague David Ignatius. I was transfixed. I couldn't even blink. They were at the Munich Security Conference talking about the nuclear negotiations, Syria, and the Israeli and Palestinian conflict.

And then suddenly, as the conversation was winding down, Ignatius said, "I'm going to take a brief point of personal and journalistic privilege. My colleague and personal friend Jason Rezaian, who's our correspondent in Iran, has been imprisoned since last July with his wife. The charges have never been made clear, and it would be wrong of me not to use this public forum to say to the Iranian foreign minister that we dearly hope—and I speak for all journalists I think around the world—for his prompt release."

Lying on my bed of rough blankets atop a rickety iron cot, I suddenly felt very tall.

You can get through this.

I thought back to my family's darkest days and how we pulled ourselves back from the edge. Even then I was sure we'd survive, I just didn't know how.

Still, so much remained to be resolved.

When my father died, my mom was suddenly a healthy sixty-eight-year-old widow whose first grand-child had also just died; my brother had lost our dad and his firstborn son. My sister-in-law was trying to hold it together for my surviving nephew, who had just turned three and had shared a room with his older brother every night of his young life thus far. They were each living a scenario that no one wants to imagine.

There was very little I could do to console any of them. Fortunately there was work to be done. There was too much inventory in the annex, Dad had some debts, and I had to figure out how I was going to get back to my life with Yegi in Tehran.

I'd known that this day would come. I couldn't have foreseen the circumstances, but I'd realized long before that one day my dad would die and I would be respon-sible for tying up the loose ends, and that there would be plenty of those.

For several years I would periodically ask my dad which dealers had merchandise consigned to him, who he owed money to, who owed him money, and who had any of our rugs. The week before I left for Europe we had lunch one afternoon and I got the updated lists, with him telling me, "Don't worry. When I feel better I will take care of all of it."

The more shameless characters in the business moved quickly, putting out feelers for what they wanted. Handwritten IOUs with signatures that didn't match my dad's and decade-old postdated checks—a bad Iranian-man habit—were laid in front of me. "Your father owed us $78,000, now you owe us $78,000," was the line, I think.

"No I don't," was my answer. But maybe I did. How could I know?

One crazy bastard who for years used to show up regularly at my parents' house early in the morning and pace in our driveway until my dad would come home from his morning exercise routine even had a lawyer—an Iranian one, obviously—send a letter demanding payment for a small silk rug he had supposedly consigned to my dad, with an astronomical price tag of $17,500 attached to it.

He had long ago earned the nickname Buster from Dr. Mary, who was annoyed by his annoying unannounced visits.

My dad had done so much for this clown, helping him get set up in America like he had with so many of his countrymen. The gall this guy had. He didn't realize I knew he ran a *hawala,* an old-school money-transferring system that became popular as sanctions against Iran mounted. People were getting five years in prison for doing that and he was trying to shake me down. I very easily could have picked up the phone and reported him out of my way. I knew the people to call.

Et tu, Buster? I thought.

But this is exactly what I had expected; I'd been bracing myself for it for years.

I was staying with my mom at the rental, refusing to get out of bed for the first couple of days, not able to accept the prospect that I had to go do another rug shop closing sale.

I went to see the Armenian wholesalers whose father was my dad's first local supplier. They gave me an updated list of the inventory my dad had consigned from them. About seventy pieces, a manageable number. They also presented me with a bill. I can't remember how many zeroes it had, but they were some of the few people who were on Dad's own list of actual debts.

Their head of merchandising, David, an Iranian Jew, was also on Dad's list. I talked to him, too. He also had rugs consigned at the annex, which I promised to return.

And lastly I went to Issa, my dad's best local rugman friend, a Kurdish Sufi with crazy hair whom I trusted in every way. We cried together for a few minutes, because we each recognized the place the other had in my dad's heart and he in ours.

His first bit of advice—more of a commandment, actually—was that I could not begin any kind of sale for forty days, the traditional Islamic mourning period. I knew that wasn't going to fly. I needed to get back to my life, and in forty days anyone who might be convinced to buy a rug because a guy died would have forgotten he had ever lived in the first place.

I was in a tough spot. A real advertising campaign would cost serious money that I didn't have. It was also the sort of time commitment that I wasn't willing to make.

Plenty of rug men stepped forward wanting to run a sale for me. The last thing I was looking to do right then, though, was go into business with anyone. I thought of one of my favorite pieces of Iranian wisdom: "If a partner were such a good thing, God would have had one."

I also knew from those initial debt inquiries that if I were to advertise a sale other imaginary creditors might step forward.

I sat down and thought about what I was going to do. It was April of 2011.

The first thing I did was take the large smiling framed portrait of my dad that we used at his memorial service and prop it up in the store's front window. I took a copy of the obituary that ran in the *Marin Inde-*

pendent Journal, which I'd written, and taped it to the window.

"Born in Mashhad, Iran, in 1939 he resided in Marin for 42 years. Rezaian is remembered for his generosity of spirit, willingness to help those in need, and a mischievous sense of humor. Rezaian was long involved in cultivating understanding between Americans and Iranians," it said, concluding with the awful detail that he was preceded in death "on March 4th, 2011, by beloved grandson, Walker Rezaian, age 5."

It was as much an announcement to the world as it was a reminder to myself of what we had just suffered.

I decided immediately that I was not going to do this in any typical way.

I put an ad on Craigslist that made it clear that this was a chance to get a rug, but that you had to make an appointment. Unexpectedly the first person that responded had known my mom in the 1980s. She was a woman who had interned with my mom when they were both beginning their careers as psychologists. She bought several rugs, but more importantly she told some friends.

One guy who walked by the shop, a musician in a kilt—a drummer, not a bagpiper—ended up buying twenty-four rugs. Unexpected, but very welcome.

I took photos of the rugs and posted them on a Facebook page I made, called "Rezaian Persian Rugs—The Ending of an Era." I shared it with friends, who shared it on their pages. Soon I was getting orders for the rugs from that page from all over the country.

And I made deliveries throughout the Bay Area. I even started what I called the Rug Mobile service, throwing a bunch of rugs in my dad's old 2000 Mercedes M-Class SUV. I'd post an ad on Craigslist, update the Facebook page I'd set up, and tweet a location for the day and then just show up in different neighborhoods and hope for the best.

I was a selling machine. No reasonable offer was refused, because I understood that my primary goal was not to get the most that I could for each rug, but to turn thousands of handmade treasures into as much cash as I could, as quickly as I could.

This was actually my second stint as a rug dealer. During the worst of the economic downturn in 2008, I'd tried out my own rug business. I called it Rug Jones. It was a disaster, to put it lightly, and I'd closed up shop after just one painful year.

Now I was suddenly the successful merchant that I never could become with Rug Jones. At no point, though, did I feel pulled to get back in permanently. It worked because I knew it wasn't forever. During those months I took a deep breath and remembered that covering news from Iran was also not the only thing in the world. That line about "the best-laid plans" is a good one. My new mind-set was to roll with life's twists and turns and to keep those I loved close.

So I offered my mom the option of coming to live with me in Iran, and she jumped at it. When she and I finally took off in August of 2011 our future was completely uncertain.

I no longer had an apartment in Tehran. Yegi had packed up my few belongings and vacated the old place soon after my dad died when the lease had ended. Mom and I couch-surfed on opposite ends of the city until I found a suitable home for us: a two-bedroom apartment in a small building close to Vanak Square in the heart of the capital, butting up against Hemmat Expressway, a couple of hundred steps from the city's most essential bus line. In every way it was a step up from the old place.

This could work for a while, I convinced myself.

And it did.

My friends welcomed my mom into our social life. When I'd have people over she wouldn't hide in her room, she would just hang out. Often she'd get invited to gatherings without me. We were an oddity, but one that people enjoyed being around. Against the tough odds of the circumstances, we had landed on our feet, she and I, right where we wanted to, far away from anything conventional and the pain of our American life.

I was proud of us for not folding. Especially her. But it was obvious from early on that the arrangement couldn't last.

For one thing the toxic mix of low levels of oxygen and gasoline exhaust that passed as our air was a major health hazard. Tehran had been one of the world's most polluted big cities already and that phenomenon had gotten dramatically worse in the year or so since sanctions over Iran's nuclear program had blocked Iran from importing gasoline, forcing the country to rely on nonstandardized fuel produced in its petrochemical factories. Just breathing would eventually kill us.

But we weren't worrying about the air just then. We were battered and licking our wounds. The fact that we were still standing and far from the many friends and relatives back in California who just wanted to hug away our pain—or was it theirs?—was a victory for both of us.

Besides, there wasn't time to think about that. We had a home to create.

I had shipped a few antique rugs back to Iran that

needed repair work best done in Tehran's bazaar. Those were our first home furnishings. Yegi called her family's longtime satellite man, Hooshie, to install a dish and receiver for us, so we could live like everyone else with a full range of forbidden channels beamed into our living room from around the world (except Tehran; we never watched local TV). I bought an old orange living room set from a friend whose dad had also just died and who was selling his belongings. I got a couple of other rugs from him, too. *You can never have too many rugs,* I thought.

During the first couple of years living in Tehran I had deprived myself in an effort to save. Now it was time to be comfortable.

The woman I hired to come two days a week to shop, cook, and clean, became an indispensable part of our existence and a trusted friend.

The autumn was a time of transition for my mom and me. Our rooms were next to each other, each with a small twin bed that had been handed down to us from one of my dad's cousins who lived in Tehran. Sometimes I could hear her crying through the wall, and I'm sure she could hear me. We usually left each other alone, but if it went for what felt like too long there would be a knock and a "Hey, you okay?" followed by a knowing shrug and sometimes a hug. We were both just letting it out. It's what had to happen. There wasn't much else to be done. We had decided we weren't just going to survive; we'd live.

I was getting back to work while trying to keep Mom occupied. Of all of the possible outcomes this was the one that suited us both.

Yadoallah motioned to me to take a walk outside so the bugs in the room wouldn't pick up what he wanted to say. I followed him into the chilly afternoon air.

"You need to stay strong. You're getting out of here. I have no idea when, but America wants you out. They won't leave you behind. These bastards, they want you to break. Don't give them the satisfaction of seeing you suffer."

I had heard this sort of thing from Iranians for years. The idea that America is all-powerful and its will hovers over everything in the world. I knew something else to be true. But I desperately hoped that he was right.

Two days later, I was told that Yegi would be paying me a visit.

"Your wife will come tomorrow," Kazem reported, dejectedly. "And she will be allowed to come once a week to see you."

This was a side of him I didn't know: conciliatory, because he was told he had to be.

"Why now?" I knew the answer but wanted his version.

"It is a gift from the Great Judge, for twenty-two Bahman." February 11, the anniversary of the founding of Iran's Islamic Republic and the most important revolutionary holiday.

That night I walked around the yard, pacing in anticipation of reuniting with my wife. Iran celebrates on the eve of holidays and over the tall brick wall and cypress trees I could see the color of the horizon changing with bursts from the outer edges of a fireworks show for the Islamic Republic's half-assed answer to the Fourth of July.

"*Allahu akbar*"—*God is great*—I heard people calling out from the public ward of the prison, physically only a few yards away from us, but another planet entirely. In between were several high brick walls, the prison's execution square, and the imaginary force field erected by the Revolutionary Guard.

I went to bed feeling hopeful about my meeting with my wife. In fact it was the first contact Yegi and I had had in two months. A conjugal visit; one of the many hard-to-believe-it's-on-the-books rights of an Iranian prisoner I had been denied in my first seven months in Evin.

But there are no coincidences or momentary re-
prieves in these cases. The pressure was working. The
IRGC had possession of me, but suddenly, due to the
growing attention, I seemed a lot more valuable. And a
liability if anything were to happen to me.

Yegi and I met in a chilly room. We had four hours
alone. It was the longest time we had spent together
since our arrest six months earlier.

When I returned from my time with Yegi there was
a lot of commotion. Yadoallah was frantically gathering
his possessions.

"What's happening?"

"They're transferring me to the public ward. One
step closer to freedom!" He was ecstatic. After two and
a half years in isolated purgatory he was returning to
the land of the semiliving. He knew he wouldn't be
free any time soon, but it was what he had most wanted
for months.

I was happy for him. He needed to get out of there.
But I was worried for myself. Who were they going to
stick me with now?

By that night Mirsani was back with me in the cell.
See the camel. No you don't.

On the one hand I was sad that Yadoallah was gone,
but on the other I felt relieved that I wouldn't have to

get to know a new guy. The odds of a third consecutive cellmate being someone I could live with seemed incredibly low.

I kept my bed and he took Yadoallah's. And that was that.

As 2011 came to a close, I had heard that Christopher Hitchens's health was worsening. He had told me when we last met the year before that his odds of survival weren't really odds at all, as the chances were so slim. On December 16 I read the news that he'd died the day before in America. There were a lot of obituaries already, because those things are prepared in advance. I knew what I thought about him and had stories to tell, but I was so far away.

I sat and cried one more time. My mom walked in wondering what was going on.

"Hitchens died," I told her.

"Oh, honey." She and my dad had met him, too, and she knew that without his influence we wouldn't have been in Tehran.

I got a text message that day from my friend Marc, an Agence France-Presse journalist. "Sorry to hear about your pal Hitchens. You've had a lot of loss this year." *The understatement of the millennium,* I thought.

Now it was time to get on with things.

Mom and I ended that year with a Tehran Christmas party. We didn't have a tree or lights, but we found some festive napkins. Friends came through. Iranian and foreign ones. Mostly journalists, a few diplomats, and a handful of locals. It was fun.

The Iranian women among them, led by Yegi, wanted to dance. Mom and I weren't going to get in their way. I can't remember ever having dancing on Christmas. We didn't have a speaker, but someone had a portable one in their bag.

We ate and drank, and at the end of the most calamitous year, we were calm.

10
Waiting to Go on Trial

Mirsani and I had both come a long way since solitary. We were in a better place now, able to at least have some contact with the outside world. The typical day started with the opening of the lock that had kept us indoors the night before. I imagine that all prison doors everywhere are the same: heavy and squeaky, because they're in need of oiling. It didn't matter what the weather was like, we always preferred to be outside.

I woke up earlier than Mirsani, usually by a couple of hours. I'd make a coffee, then I'd go out in the courtyard and do some hundreds.

We knew all too well that we were bargaining chips being held by reckless poker players who were over-representing what they had in their hands. They were like rug merchants trying to sell product at a thousand percent markup. I knew this game, and I just wondered if American negotiators did, too.

It was talked about openly. Zarif would say how offensive such references were to the art that he practiced, but the only art this guy knew was the fine art of bullshit.

But this reality, that we had value, manifested itself in one very important way. Food.

A hand-me-down from Yadoallah—an electric cooking pad—became an extra limb for us. And we got proficient with it.

Apparently in a lot of places, including Iran, if you've got money you can get stuff in prison, and I don't mean smuggled in *Shawshank* style. There's an actual system.

We didn't have access to the prison's commissary, but once a week we were given a form with six spaces and the amount of money we had to spend, based on what our families left with prison officials. Every week we were pushing the envelope when it came time to do our shopping lists and some weeks we failed, depending on who the guard was on the day to fill out our forms; neither one of us could write in Farsi so we always needed help. Three kilos of ground lamb, forty eggs, ten kilos of potatoes. Believe me when I tell you that not having the freedom to move is tough regardless of the circumstances, and for that reason you look for any way to make it easier on yourself. For Mirsani and me it was through our stomachs.

The fact that they were allowing us decent food further reinforced for us the idea that we weren't regular prisoners.

As the months wore on we took this one right and turned it into something that no one at Evin had ever seen before.

We had another good thing going: visits from Mirsani's mother, Elmira, and the home-cooked meals his wife would send. I learned more about Azerbaijan from those care packages than I ever could from any book.

At his home he must have had a lot of land, because every visit would bring fresh tomatoes, peppers, and eggs, and twice they brought whole chickens, raised in his backyard.

They would send him baked bread and cookies and preserves that Mirsani's wife prepared.

We figured out over time that sweets, for all their teeth-corroding and diabetes-inducing properties, are one of the keys to short-term bliss. Maybe you're wondering, *Wasn't this guy ever a kid?* I most certainly was and I loved candy, but not as much as I did while I was in prison.

Sugary treats were so important we began to plan around them. First it was our shopping lists that needed to take our sweet tooths into account. Then it was how we organized our days. Neither Mirsani nor I wanted to be the middle-aged guy eating chocolate for breakfast. That's just not a good look.

Whenever Mirsani's mom was arranging to visit we would do everything we could to figure out who would be the guard on duty that day and whether or not they would be complicit in getting the booty into the prison.

Sometimes we failed and boxes filled with delectable items—stuffed grape leaves; homegrown chickens; sugar cookies with a walnut filling; mille-feuille, the flaky layered pastry with vanilla custard known in that part of the world as a napoleon, which Mirsani's wife made better than any other I've ever tasted—were blocked at the prison gate and sent home to rot with his mom. These were proud people and if the man of the house couldn't have good things to eat while he was in prison, his family wouldn't either, for a few days.

After a while he impressed upon the guards that, even if they wouldn't let us have the food, at least they could tell his mom that it had been accepted. They agreed to that. They were all assholes, but they all had moms, too.

On those days we sulked, but only until Mirsani

would make his short call back to Jolfa to his wife. He had to put on a good face. She would want to know how we liked what she had sent, because as Mirsani always told me, "She knows she's cooking for two."

My situation seemed to be stabilizing, not that it provided much hope. In the corner of our yard there was an old fountain covered in dirt with some gnarled plants sticking out. Roses started to bloom and I began, timidly at first, using a contraband knife an earlier resident had left behind, to cut one or two to take with me to my visits with Yegi that had just been reinstated. We were able to see each other on Tuesday mornings for an hour and were given sporadic conjugal visits when Islamic law stipulated that it couldn't be ignored any longer. I wanted to do anything I could to manufacture some sense of normalcy and beauty during such an ugly time. I never got any blowback from the guards. By that point, despite whatever danger they were told I supposedly posed, they knew I was harmless.

It was a tough high-wire act they were trying to pull off: hold me as a foreign hostage, but treat me as an Iranian prisoner.

When she had been freed on bail Yegi was forced to sign a contract saying that she wouldn't speak to the media, but threatening to publicize the denial of our religious rights as husband and wife, at a time when world leaders were beginning to take my detention seriously, was the only leverage she had over the court and IRGC, and she used it skillfully.

When Yegi visited me at Evin Prison on March 14, 2015, she smiled in a way I had not seen since before our arrest nearly eight months earlier.

"Muhammad Ali issued a statement calling for your release," she said, beaming.

Initially I thought she was just trying to lift my spirits. I had told her recently that I did not want to hear any more bad news about my situation, which looked hopeless.

It was the day before my thirty-ninth birthday and I was at a low point, suffering from the weight of a long forced isolation, but once she convinced me that it was indeed true, I cried the happiest tears of my captivity and felt the strongest I had in months.

It was a turning point for me. The public acknowledgment by Muhammad Ali, one of the most unifying figures in the world, that he believed I was innocent of any wrongdoing meant everything to me.

And that blow had real impact.

We were turning some kind of corner, but the road had so many twists and turns ahead. I didn't know it yet, but the statement was part of the efforts by the other Ali in my life, my big brother, and his new and well-positioned friends in Washington, to draw attention to my plight.

After the Greatest of All Time released the statement on my behalf, several of my prison guards told me they had heard about it, and some began to treat me differently—better, and with more respect. I like to think that his words made them doubt the forces who signed their paychecks.

The next day in the cell, flipping through the channels, I chanced on a documentary about Muhammad Ali and his legal battle against being sent to fight in Vietnam. I knew people around Iran would be watching it and some of them would have heard about his statement in support of me. I am one of literally billions of people affected by the life and actions of Muhammad Ali.

At the time, a domestic Iranian news agency wrote, "One of the most astounding moves by the U.S. government and Rezaian's family was bringing the famed American boxing legend Muhammad Ali, who is very popular in Iran, into Jason's freedom campaign. They used his popularity to influence Iranian public opinion."

There was an odd truth to this logic, but not the one they intended.

I needed that shot of strength, as the legal part of my ordeal was just getting started.

I don't think I'm ruining the ending by telling you that if you're put on trial in a court that has "Revolutionary" in its name, you shouldn't expect to win.

Anyone living in Iran or covering it closely has some awareness of Abolqasem Salavati. He is a judge in Tehran at Branch 15 of Iran's Revolutionary Court.

Salavati, often called the "Judge of Death," for his affection for sentencing people to death by hanging, was chosen to hear my case. Throughout my trial and the two preliminary meetings I had with him before the trial began, I was able to clarify that I, without the benefit of any legal training nor access to legal representation, seemed to know much more about the law than he did. All he knew was that whatever he decided, or was compelled to decide, was law. I tended to disagree and I let him know it, backing it up with arguments. Ours was a strange and strained relationship.

My case was taking place inside a very small ideological vacuum, one that I knew would not be impacted by whatever I did or said. Plea or no plea, I was guilty and the sentence was already written. All my team, which consisted of my wife, my mother, and local lawyers, could do was get as much information as possible outside of that bubble. The further away from the inside it got, the more absurd it would appear. Proving my innocence to the world, and thus continuing to receive support from far and wide, was the only defense available to me.

Farcical institutions require cartoonish figures to front them and Salavati fit the bill perfectly.

Not a lawyer with a polished educational pedigree. Not someone who had particularly well-known revolutionary bona fides. Not even a cleric. Just a scary-looking guy who most people agreed had been a janitor at some point, and now held the key to the courtroom where Iran's most high-profile cases—especially ones that involved foreigners—were tried.

Mine was a case of the highest crimes and only Salavati was experienced enough to hear it, the narrative went. He could be a punch line to every tasteless joke about all that is wrong with the Islamic Republic.

Salavati, it turned out, was the judge in the case of every single captive in Evin 2A, including both of my cellmates.

"He's a butcher," Yadoallah had said.

"He's that evil?" I asked, scared.

"No," he guffawed. "Like a butcher everything is done by weight. He hands out sentences by the kilo." That assessment stayed with me in the months that I waited to finally be called before this mythical figure.

I was feeling stronger. I was spending time in the gym, walking for hours in circles in our yard, reading, and my sporadic access to calories kept my weight down.

More than anything, though, it helped that the intensity around efforts to get me out was undoubtedly picking up, rising to a climax at the White House Correspondents' Dinner on April 25, 2015. I didn't see it, but this is what Obama said right after he finished his annual stand-up routine.

"Now that I got that off my chest—you know, investigative journalism, explanatory journalism, journalism that exposes corruption and justice, gives voice to the different and the marginalized, the voiceless—that's power. It's a privilege. It's as important to America's trajectory, to our values, our ideals, than anything that we could do in elected office," Obama said.

"We remember the journalists unjustly imprisoned around the world, including our own Jason Rezaian. For nine months, Jason has been imprisoned in Tehran for nothing more than writing about the hopes and the fears of the Iranian people, carrying their stories to the readers of the *Washington Post,* in an effort to bridge our common humanity. As was already mentioned, Jason's brother, Ali, is here tonight and I have told him personally, we will not rest until we bring him home to his family safe and sound."

Obama wasn't really given a choice that night. He had to talk about me.

My brother, the *Washington Post,* and WilmerHale, the law firm they hired to help build a case against Iran, along with the community of organizations fighting for journalists' protection and many of my colleagues, had made sure of that. Besides writing op-eds and mentioning me in their remarks, many of them distributed and wore #FREEJASON pins.

The lift I got from hearing about it continued the wave of confidence that Muhammad Ali had inspired a few weeks earlier. I needed it as we entered into a new season of my ordeal.

A couple of days later, back on U.S. soil, where he's always most comfortable, Zarif was at it again, talking to David Ignatius, this time at NYU, just a few blocks away from the New School and my old stomping grounds.

"Mr. Minister, I want to ask you one more question. And it's a personal one, because it involves my colleague Jason Rezaian, who has been imprisoned in Iran on charges of espionage that his family, his newspaper, and now the U.S. government, in the voice of President Obama last Saturday, say are false," Ignatius began. "In the spirit of the moment, we're talking about momentous agreements, in the spirit of what President Obama's called mutual interest and mutual respect, wouldn't this be a good time for the release of my colleague Jason?" Ignatius asked.

And then Zarif, this green-card-holding, U.S.-law-degree-wielding sonofabitch of a foreign minister of a sovereign state, understanding full well how slander in the United States of America works, said:

"Well, as I told you in Munich and I'm telling you again, that I hope that no one—nobody will be lingering in prison, including a lot of Iranians who committed no crime across the world but are waiting in prison to be extradited to the United States for violating U.S. sanctions, which are illegal anyway. One of them died in the Philippines in prison. So I'm not trying to make it quid pro quo, but I'm just saying that, of course—I mean, the *Washington Post* has a much better publicity campaign about Jason than we have about people who are lingering in prisons in Southeast Asia as well, who committed no crime. Unfortunately, your friend and my friend, Jason, is accused of a very serious offense. And I hope that he's cleared in the court. But he will have to face a court. He's an Iranian citizen."

Jason Rezaian, your friend and mine. That's probably what they'll put on my tombstone.

As much as this back-and-forth angered me in ways that blurred my vision, I was still able to see that it was creating buzz. Even the Iranian propaganda machine wanted to make sense of what the struggle over me really meant.

The very next day I was watching Press TV and one of my favorite shows, *Face to Face,* came on. The basic premise is the same as that of every Sunday morning newsmaker show: a one-on-one interview, but in this case almost always a minor official or visiting foreigner to Tehran, with the channel's director of news, the clownlike Hamid Reza Emadi. On this particular episode, he was talking with Elham Aminzadeh, President Rouhani's assistant for citizen's rights.

"If the United States realized one of its nationals is in trouble in another country or one of its nationals is coming under any kind of restrictions by another country, the government comes to that person's help. And even we have seen that if an American national gets killed anywhere in the world the U.S. government takes an official position. Or if an American national gets jailed in another country, like *this person* Jason Rezaian, who is in jail in Iran being investigated for things that the Iranian judiciary is probing right now, the U.S. government takes an official position. We don't see that much support for Iranian citizens on the part of the Iranian government. Why do you think that is so?" Emadi asked, in one of the many instances where a regime devotee unknowingly undermined his own position by asking a poorly formulated question.

"This is the weakness, actually, of our organs to implement diplomatic protections of nationals, Iranian nationals abroad," was the only feeble, albeit honest, answer that Aminzadeh could muster.

I remembered the one question that Iranians with no read on the outside world would invariably ask me and other dual nationals. Usually a taxi driver or a college kid. "Which one is better, America or Iran?" It was an uninformed question that felt impossible to answer. "Both of them have their pros and cons," had been my stock reply for years.

But now I had the undeniable answer.

The person chosen to represent the rights of Iranian citizens, appearing on Iran's international propaganda channel, was admitting to me, probably the only person in the world watching and one of the few whom it mattered to, that Americans have rights in the world and Iranians don't and that her government couldn't and *wouldn't* do anything about it.

By then I had a prison routine. I was on a kind of

autopilot, doing time—"drinking cold water" as they say in Farsi—with no end in sight. I had the gym and my books, and there was no new reason for my captors to cause me any fresh suffering, so they mostly left me alone.

Tuesday mornings had become the centerpiece of my rituals. I was led blindfolded by a guard out of my cell, down the open-air corridor that flanked it, up a flight of brick stairs, and into a short indoor hallway past the warden's office.

Sometimes someone would talk to me. "Mr. J, how are you this morning?" If it was quiet they would be friendly. "You're still here?" It was the sickly passive-aggressive and uniquely Iranian way of expressing oneself in an uncomfortable situation.

The awkwardness of walking blindfolded never goes away and sometimes I would bump into a wall if a guard wasn't guiding me, at which point I would call out, "My human rights! What about my human rights?" Everyone would laugh.

For several months I refused to shave.

"What's up with the ISIS beard, J?" someone would ask.

"I'm in mourning," I'd reply, deadpan. Shias go unshaven when they grieve.

"What are you mourning?" they asked.

"Justice in the Islamic Republic."

They'd just let it go.

I had been there long enough that the prison staff was used to me. Everyone knew the score by then. I was, like so many former residents, just waiting to be leveraged.

Other times, if it was busy and there were interrogators or agents working other cases milling around, the attitude was colder. Conversations were carried on in hushed tones and the corridors smelled more strongly of unbathed Middle Eastern men. I hated those people, although I never saw any part of them besides their feet, usually wearing cheap rubber flip-flops, similar to the ones I was given to wear.

The same guard who the day before might have been on a first-initial basis with me—and probably would be again an hour later—would say, "Sixty-two, pull your blindfold lower and keep your head down."

On Tuesday mornings, passing the office door, and in open air again, my excitement and anticipation would grow. This was my life. An hour, once a week, behind a glass window with my wife. I had absolutely nothing else to look forward to.

Some weeks there was no news. Nothing. Other times it was bad: a hardline MP had called for my execution or my lawyer had been denied another request for rights that were legally available to me. I had gotten used to it by the spring of 2015. The fix was in and everyone had a role to play, including me. Mine was to survive.

It was during a visit in early May 2015 when my mom made her first appearance behind that window.

I hadn't seen or spoken to her since the Christmas encounter. I was glad to see her but also felt more hopeless than ever. She'd returned to Tehran because my lawyer had advised that my trial would be starting "soon." I didn't believe that, but I knew that when Mom was in town there was a greater sense of urgency: more international news stories and more pressure on her from my captors to "get Obama and Kerry to do something."

" Everyone knows what an injustice this is," she assured me. "We are working to get you out of here."

"Work harder," I said with a sarcastic half grin that she knew better than anyone.

There was so much I wanted to ask her, but as usual I forgot the most important questions. I wasn't allowed to take anything, especially not written notes, with me to those weekly meetings. We talked about friends and family, and I tried to jog my memory but couldn't.

I was just looking for a positive sliver to latch on to even for a moment.

"Hey, Mom, is it true that Muhammad Ali called for my release?" I asked her. I'd believed it when Yegi had told me that around my birthday, about two months earlier, but now it seemed so long ago that I might have imagined it.

"Of course," my mom replied in her animated way. "That was really something."

"Yeah." I pondered it for a moment. "So you're telling me I've got the two most influential American Muslims on my side?"

"Muhammad Ali and Congressman Keith Ellison?" she said.

"No." I paused and smirked. "The Champ and Obama."

I knew what I desperately needed was for there to be more color to the story.

Of course there was nothing I could actually do from inside and only a limited amount I could communicate to the outside world through my contact with Yegi and my mom.

"Get people to say whatever they want about me as long as it's true," I commanded the two women in my life.

My belief was that the public was drawn to compelling figures, making them even more attractive to the news business. With the reach of social media soaring, I believed that people could help drive the conversation. And the longer and more often I was talked about—and the more diverse the audience who heard about me—the more likely I would get home.

As I knew from my very cursory readings of old online marketing books when I was still selling rugs, it's important to keep your business in the public conversation, but it's almost equally essential to maintain honesty and transparency while doing it. People smell a rat very quickly and once you've lost them it's hard to get their trust back.

"Write to everyone who knows me and ask them to post whatever they know about me on their Facebook pages," was a command I gave to my mom during a meeting that summer.

It may have sounded simple or even dumb, but I understood that in this age if everyday people don't care, no one on the top will either. It's easier than ever for normal people to make their concerns known.

By then the outpouring of public support for my release—and against Iran's egregious actions—had hit high decibel levels. More than half a million people had signed a petition on Change.org demanding my freedom. Well-recognized figures from politics, media, entertainment, and the scholarly community had all made very public calls for my release. Critics of the American system—the sort that the Islamic Republic considers their allies—were very publicly taking my side.

Edward Snowden tweeted, "Iran's shocking conviction of a journalist on secret evidence must not stand. #FreeJason."

Noam Chomsky and two dozen other well-known academics published an open letter demanding my release and calling my trial a sham.

"No matter what you do to me," I reminded Kazem every chance I got, "you can't win."

What could still make a difference, I believed, were individualized statements on my behalf by people who knew me. Testimonials. I didn't have the right to call witnesses in the Revolutionary Court, but I knew there were people I had met all over the world who were ready to testify on my behalf.

Amber Thorsen, my first love and a lifelong friend, took it upon herself to tell people about me. She did so at work and on the Internet, and she wore a T-shirt—made for her by a friend—that she called her "human billboard" that read "Free Jason & Yegi" over "Change.org."

Several friends in different parts of the world, unaware of each other and in solidarity with me, didn't shave for months.

It was impossible to measure the impact of this outpouring of love and support on decision makers in Tehran, but I knew that it would matter in Washington and New York, where the conversation was being led in the media.

And more than that, hearing about all the ways I *hadn't* been forgotten gave me much-needed periodic shots of hope. It renewed some of my lost faith in humanity and reminded me of one of the great things about Americans, born from the freedoms that have been endowed to us. This is one of those arguments that no Iranian nationalist or anti-American anywhere hell-bent on painting America as a force of no good will ever be able to win. American lives matter. (If enough people give a shit.)

I was not so naive as to think that the voice of the people was going to shake me out of Iran. Self-proclaimed experts are torn on the idea of how best to engage the Iranian regime on hostages they've taken. Some say it is essential to make as much noise publicly as possible, that the regime only responds to pressure. There is anecdotal evidence to suggest that may be true.

Others believe that the regime in Tehran is concerned with its public image and for that reason anyone who attempts to sully its name any further risks raising its ire, which leads authorities to take capricious action. That everything is resolved in due time.

I would offer an alternative view. No authority in Iran acts—in this case takes a hostage—without seeking a reward. It would be giving them too much credit to think that at the point of abduction the power involved knows what it actually wants in return. It would also be a mistake to believe that other actors in the system will wholeheartedly support this action. No, there is a lot of internal debate over the merits of taking a hostage, the pros and the cons. The short- and long-term effects at home and abroad.

It would be yet another mistake to think that simply talking can help get hostages out. *Reason with them and they'll see the light,* goes the thinking. *Find a face-saving way to defuse an embarrassing situation.* Here's the thing: If you're worried about saving face you don't take hostages. Period.

All of these beliefs undercut a core value of Iranian mercantilism, which is the basis for its diplomacy. It's the old tenet "Yeah, yeah, yeah, but what can I get for it?"

Without testing the market and setting a price, Iranians won't let something that's fallen in their lap go. It just doesn't work like that.

And when the value finally gets set, the initial price is going to be very high. So best to dig in, sit back with a smile on your face, enjoy many cups of tea, and realize you may not even end up buying a rug from this guy after a lot of back-and-forth. And if you don't it's not the end of the world. Although they're each unique and made by hand, there will always be others.

Trading for rugs and hostages, in turns out, is pretty much one and the same.

It may have been reckless to want my perceived value to be raised. America knew I hadn't done anything wrong and so did Iran. But without a significant campaign to raise awareness nothing was going to happen.

Call it a savvy acquisition or dumb luck, but the intelligence wing of the IRGC had this working for them throughout my imprisonment.

On the outside, my brother continued to crisscross the world. He made twenty trips to Washington in 2015. He was away from home for over two hundred nights that year. He met with heads of state, arms dealers, journalists, and diplomats. He was on regular phone calls with the *Washington Post*'s top brass to discuss progress. Extraction plans were floated but were later deemed too risky. There were discussions about running a Free Jason ad during the Super Bowl.

Jeff Bezos worried that there was a real chance that I'd be executed.

While all of that was going on, it became crystal clear to me that I wasn't coming home until Barack Obama decided I had to.

As the time dragged on, there were flurries of activity and then weeks, sometimes months of nothing. It was hard to decide which one was worse.

During the nothingness I just craved the presence of someone who could answer the endless questions running through my mind. I knew if such a person came along their replies would be utter bullshit, but by the spring of 2015 I was better equipped to interpret it.

On those days when there was action, though, I just wanted to be alone and not have to interact with the willfully ignorant and blatant hypocrites that were my captors and the structures that kept them employed.

Since Christmas Kazem and Borzou had been telling me that my trial would be starting "next week." The feeling of anticipation and fear was excruciating.

On the one hand, I knew I was going to lose, and on the other I knew that until I lost there wasn't even the slightest chance that I would be set free. You can call it Orwellian or a catch-22, but to me it was just a well-established fact that I wouldn't be released until they hung a stiff sentence on me.

When looking at the cityscape of Tehran one sees that it is littered with unfinished construction jobs. Half-completed eyesores that gather dust over decades. They make up for it with what they consider their judicial efficiency. It's very rare that a trial does not end. "Justice," the Koran dictates, must be served.

By the time my trial began on May 25, 2014, I already knew that it was a date to look forward to, an obstacle that had to be overcome.

My captors argued that I was attempting—through my reporting—to change American behavior toward Iran and that this was my crime. If so, guilty as charged. Perhaps. Everyone was beginning to see that the case they were making against me would be a very hard one to sell in court if I decided to defend myself.

But on the surface that is what this case was all about. Take a guy who is, with some level of empathy toward Iran, describing in plain English the various elements of the Islamic Republic's ethos to an audience that for decades decided it did not want to or could not understand it. It was too much for hardline ideologues to comprehend, which means it was impossible for them to accept. If you can't own it, control it, or understand it you must destroy it. That was the attitude I found myself up against.

And when it came time for my day in court I wondered if the judge's understanding of the world and the nuances that flourish naturally in the light of day would have a wider range of colors than that of my interrogators. But it became very obvious very quickly that my judge, Abolqasem Salavati, was, without exaggeration, one of the dumbest people I have ever encountered.

As usual, in our last meeting before my trial started Yegi had some good intelligence and knew my first session was coming in a matter of days. "You will go to court and you will fight," she commanded. "There is no pleading guilty."

I nodded.

"Answer the judge's questions completely, and in each case add that 'this act is not a crime,' but keep your answers short. Less is more." My wife's English was excellent when we met, but I'll take some credit for her expert use of American idioms. I felt lucky and proud to have her as my only link to the world.

"I got it."

Those were my only marching orders.

11

The Trial

May 25, 2015

TRIAL SESSION 1

On the day of the first trial session I was told at eight A.M. to come to my cell's gate. Despite the heads-up from Yegi that the trial would be starting that week, I'd had no way of verifying it.

I put on my silver Adidas sweat suit. I knew from previous excursions outside the prison walls that they would force me to put on my prison clothes, to brand me in the public's eyes as a criminal, so I hid in our toilet stall to waste time until the guard started yelling that I was late.

When they took me to the warden's office someone called out that I needed to change my clothes, but I just kept saying, "My expert said I could wear whatever I want."

That wasn't true. Kazem would never say that. But I kept saying it anyway. The warden, whom I rarely saw, said there was no time, and I claimed my first tiny victory: the freedom to wear my own clothes.

That day it was the same routine as I had been through every other time I left the prison. I was blindfolded and put in the back of an ambulance with curtained tinted windows.

After a couple of minutes on the freeway, I was allowed to remove the blindfold. I recognized all the guys along for the ride, including the driver, that jovial prick.

When we arrived at Branch 15 of the Revolutionary Court, we entered through a back gate. Normally anyone who was packing heat had to relinquish their arms when entering a government building—anyone, that is, except the IRGC. But the Revolutionary Court was an exception, and there was a minor argument at the front gate between my chaperones and the court's security people.

I had learned that these guys adored their guns and forcing them to hand them over was the most emasculating thing that could be done to them. I loved watching it happen.

It turned out that it didn't matter what I was wearing, because no members of the press or the public were allowed to see me. I knew from writing about other cases that journalists—domestic and foreign ones—gathered in front of the court on major trial days to catch a glimpse of defendants. *Are they in the prison clothes? How much weight have they lost? Any visible bruises or scars?* As much as I hated the idea of being the focus of such a spectacle, there was nothing more that I wanted right then than to be seen by my local colleagues. *This could be any one of you,* I longed to tell them.

We parked in the back of the building and I realized that there were two other cars along with us. Every time I left Evin I had an entourage of at least ten armed clowns guarding me.

At the main gate buses filled with dozens of inmates waiting to be tried were unloading. Most of the passengers were supposed drug offenders, shackled together and marched into the building single file. Even with the deck stacked against me I knew I was in far better shape than those guys. Officially, Iran's judiciary carried out 977 executions in 2015.

I paced a little but couldn't go far, and then I was led up the stairs into an office building that looked exactly like every other center of Islamic Republic bureaucracy. Long hallways with fluorescent lights, lots of closed doors with people hopelessly waiting outside them. Signs extolling the virtues of the hijab and others threatening women who didn't wear one properly. I had wasted so many hours in this sort of bureaucratic office over the years—trying to get my military exemption (which is sold to citizens born or raised abroad whenever the Islamic Republic needs cash), or my press pass, or my dad's death certified—where nothing ever happens according to official procedure. I knew this place.

I told my guards that I wanted to go to the bathroom before being taken into the courtroom and we wound around the large off-white hallways, passing nothing but closed doors until we found a men's room. One of them accompanied me in and stood by the door of the stall while I did my business.

Coming out I could see turnstiles and security gates in front of me and realized it was the court's front entrance. Blocked from entering, behind the gates were my mom and Yegi. I threw my hands in the air and shouted, "I love you, Mom! I love you, baby!"

This caught my guards by surprise and the driver muffled my mouth with his hands. I didn't resist, but neither did I slow my stride. I knew it was the right move. At once I represented strength and love to my mom and wife, and they had a small but impactful image of suppression and abuse that they could relay to others. *Not only was Jason Rezaian not allowed any contact with his lawyer, his mother and wife were barred entry to the court and he wasn't even allowed to speak to them.* Another detail that could be included in a story of my first day in court.

My heart began to race. This was a new chapter in my saga. I was scared of the coming barrage of questions, but I was confident. By taking me out of the prison walls and expanding the sphere that my case inhabited, the IRGC was loosening their control over the situation and increasing the variables at play, providing me with a hint of influence for the first time since my arrest.

My guards were pissed by my emotional outburst.

"You don't speak to anyone without our permission. In fact, you only speak to us or the judge. No one else. Got it?" the driver said. He had driven me every time I left the prison: to buy clothes at the high-end men's shop for the purpose of my completely optional forced confessions, to the few perfunctory doctor's appointments at an IRGC hospital when I had serious health problems and was told I was completely healthy, and to the lone meeting with my attorney two months before the trial started, where he was in the room the entire time.

He was in his late forties, had a belly, employed a comb-over to disguise the deep recession currently taking place on his hairline, wore stylish eyeglasses and colorful shirts. He seemed happy. I hated him passionately. I had since that first encounter.

"I was just telling my wife and mother that I love them. Is that a crime, too?" I asked.

"They are your wife and mother. They know you love them. You don't need to scream it in the Revolutionary Court. If that happens again we will have a problem. Understood?"

"No problem," I replied. Resisting their orders would do nothing except cause me greater discomfort.

They walked me to the actual courtroom. Along the way we passed three twentysomething guys dressed in faded jeans, T-shirts, and sneakers. Wearing the clothes of today's global youth, they could have been from anywhere. They recognized me and whispered to each other. One of them smiled. I just nodded.

The guards sat me in the courtroom and left the door open to the hallway. Within a minute the three walked by, ducking their heads in and giving me the thumbs-up. I smiled. The walls were cracking. Even parts of the Iranian public were on my side. Yegi had told me so, but this was the first time I could see it. I felt my back straighten and my confidence swell.

I learned later that those three young men were musicians who faced charges of distributing underground music. They lost their trials and started serving three-year sentences in June 2016.

One of the guards slammed the door shut.

I looked around and besides the emblem of Iran's judiciary—to me always a farcically sinister-looking take on the international symbol of justice: a balanced scale with a sword dividing it down the middle—the courtroom looked like a waiting room in any Islamic Republic bureaucratic office and probably was used as one on days when court was out of session.

The chamber began to fill up with a lot of people whom I didn't recognize. The only ones I had ever seen before were my lawyer; the prosecutor; the court clerks, each of whom came to greet me personally; and my transporters. Behind me there were at least twenty other people, all of them men with facial hair of varying lengths.

Next to the judge's elevated bench were two television cameras pointed directly at me.

When Salavati entered the chamber, I asked if I could approach the bench. I stuck out my hand and he shook it. Nodding his head toward the cameras he said, "Look at what a headache you've made for me."

"Headache for you? I have been in prison for a

year for doing absolutely nothing and the whole world knows it," I reminded him. "*Hajjagha,* you have the power to end this whole story now and send me home." I knew that was only partly—though officially—true, but I said it anyway. He didn't respond and I was led gently by the arm, by the bailiff, to my seat.

Salavati brought the room to order and read from a script what I was accused of, the case number, and a standard explanation of everyone present's responsibilities and my rights; the requisite song and dance to give the proceedings an air of being official if not authentic.

One of the clerks rose and read a passage from the Koran. That, it seemed, was his only job. I learned later it was the passage that is always read for people destined for the gallows, the fate of traitors. It's designed to intimidate, but I don't speak Arabic.

When he was finished Salavati then began to read the charges against me, but I interrupted him.

To this point, despite threatening to execute me multiple times, the "Hanging Judge" had been completely cordial in our short series of interactions, but now the courtroom was full and the cameras were on.

"Your Honor, before we begin I would like to make a couple of requests and ask a few questions with your permission," I said. "If I have that right," I added in for good measure.

"But make it quick."

"I would like to request that I be given the opportunity to be released on bail."

"Denied."

"I would like to know who all of these people in the room are," I said, motioning mostly to the people behind me.

"You want to know who they are? Why does it matter?"

"I was told this was a classified session. It looks very crowded."

"Don't worry, they all belong here. They are with us."

"With all due respect, how do I know that? And who is *us*?" I asked, subtly trying to remind him that, even if he wasn't, he was supposed to act as though he was impartial.

"Because I am the judge, and no one comes here without my permission." He had denied several of the lawyers my family chose to represent me, citing the same rationale.

"I would like to request that we delay this hearing until my family, a representative of the *Washington Post,* and one from the Swiss embassy are in attendance."

"Denied." He was starting to get testy.

"I would like to request that this be an open trial. If I am guilty of any crimes the people of Iran have the right to know about it. I would like a public trial."

"I know why you want a public trial. You want to fill this courtroom with your spies," he said, accusing me like a little kid might.

"My spies?" I was genuinely shocked he'd taken that line. "I have no idea what you're talking about."

"Of course you do. We've interrogated thirty of them who work for the domestic press and report directly to you. These are their interrogation files," Salavati said, pointing to stacks of blue and pink folders in front of him.

I laughed indignantly, feeling something like strength swelling inside me for the first time in months, and gave him an eye-rolling look that said, *Come on.*

"If you have a single witness, bring them here to testify against me. Bring one. Name one."

"Enough. We're getting started. You've wasted enough of my time already."

I can hold my own in a back-and-forth with this clown, I thought, *but will I be able to handle a death sentence if he pins one on me?*

Salavati read the four charges against me. I pled not guilty to each of them.

I could hear shifting in the seats behind me. Apparently it wasn't going according to plan.

By all accounts I had ever read about other dual nationals and journalists who had been tried, their trials were incredibly short. Often just a few minutes. No evidence. No witnesses. Just a guilty verdict and a heavy sentence.

From the beginning this was a different kind of show, and one that appeared to be being produced for local television.

A man in his fifties sat down next to me and introduced himself by saying that I could rest assured that he was "totally impartial."

"Are you a member of the court?" I asked him.

"No, I'm just a translator," he replied.

"Too bad."

I had requested an official translator, although by then, ten months into living a life solely in Farsi, my language skills were good enough that I was unlikely to miss anything if I didn't have one. But I wanted that extra level of accountability on someone else's shoulders and the buffer between the judge and me.

"So how bad is my situation?"

"It is not good, but I am hopeful for you. You must be strong," he advised.

"I didn't do anything wrong and I have not been allowed to consult with a lawyer," I responded. "You can see how bad that looks, right?"

"Yes. You should have been more careful. Journalism is not a good job."

I was beginning to like him less and less.

The procedure was the same as it had been in all the pseudo-judicial proceedings I'd been through during those first ten months. Salavati would ask a question, usually a very pointed one like "As an agent of the United States government working illegally in the guise of an accredited journalist, how do you explain your interactions with known antiestablishment figures and CIA officials such as . . . ," and then he would name well-known journalists or think-tank scholars.

I was to answer in writing, in English, and the official translator would translate the text from English to Farsi and then read his translation back to the judge as my testimony. Not exactly foolproof.

By then I knew that my sole objective was to not concede the slightest hint of any wrongdoing. *They zig, you zag,* I reminded myself. That was my role in this, and tackling it became easier than I ever expected, knowing that more people around the world were standing up in my support every day.

I had also answered all of these questions so many times in interrogations in the previous ten months, and thought them through in the quiet loneliness that followed. In every instance I just went back to my original, honest answer, and bolstered it with the further explanations I came up with during so many sleepless nights.

In the vacuum of the prison's interrogation routine, ferried back and forth from solitary, the deck is stacked against you in a way that makes the game not even worth playing. But here, with people watching, a lawyer representing me—or at least trying to—a prosecutor who'd spent a year building a wobbly case against me, and cameras, I knew I could shine.

I had no other choice.

With every question I looked for the opportunity to unequivocally state my innocence while offering reasons why what I was being accused of was not a crime.

"I am a journalist, and all of my activities have been conducted as a journalist, and all were legal," was how Iran's domestic media quoted me from the supposedly secret trial later that day.

The charges against me as read by Salavati were "espionage through collecting classified information and providing it to hostile governments" and "spreading propaganda against the regime," but it was plain to everyone that what I was actually being accused of was gathering public information and sharing it with the world through the *Washington Post*. We call that journalism.

Salavati asked a pointed question about sensitive economic data that I had supposedly transmitted, referring to reports I had written about the highly cited crash of the Iranian currency as a result of Iran's being cut off from the global financial system. This was an established fact that was openly discussed by everyone in the country and all those watching it.

I wrote my answer, which my official translator was then tasked with officially translating.

"I only ever had access to and shared publicly available information," I wrote.

And then he wrote, and read his version for the judge.

"I only shared *government* information."

I had to laugh. What else was I going to do? I leaned over and put a hand on his shoulder.

"My brother, the words for 'public' and 'government' are not the same in Farsi or English," I told him.

"Yes, yes. You are right. My apologies."

"Please explain to the judge that you made a mistake," I requested.

"Your Honor, I have made a mistake." He was obviously ashamed. "I used the word 'government' when I should have used the word 'public.'"

"Since when does the defendant give orders to a

member of the court? You will do as I say, not him!"
Salavati reprimanded him.

"Don't worry about it," I leaned over again and told
him, "prison in Iran isn't that bad. Much better than
Guantánamo."

He was obviously nervous. "I'm very sorry."

"I know you are. Don't worry about it too much."
And then, trying to invoke all that I'd learned from
a preadolescence watching *L.A. Law,* I said, "Your
Honor, I'd like the record to reflect that the official
court translator is not qualified to do this job."

The translator looked frustrated.

"Please translate," I requested, and he did.

That session ended as it started. I approached the
bench and asked Salavati to release me on bail. When
that was denied I asked for more meetings with my
wife. Denied. More telephone calls. He said he'd consider it. I stuck out my hand and he shook it again,
without looking me in the eyes.

The prosecutor approached me and said, "Why are
you making this harder on yourself? Just plead guilty. We
are not trying to torture you. This isn't Guantánamo."

I started to formulate a sarcastic reply and thought
better of it.

My guards came to take me away. Too much frater-
nizing for their comfort.

As they led me away the prosecutor and Salavati walked in the other direction, together, yukking it up.

The news of my not-guilty plea broke swiftly via the semiofficial Fars News Agency—no one in Iran even flinches saying its name even though some of them must know by now that when pronounced accurately by English speakers it is "Farce" News—which carried a report with some of the details of my supposedly secret trial.

My media machine—my big brother on TV and the *Washington Post* in print—went into overdrive.

"There is no justice in this system, not an ounce of it, and yet the fate of a good, innocent man hangs in the balance," Marty Baron wrote. "Iran is making a statement about its values in its disgraceful treatment of our colleague, and it can only horrify the world community."

Ali and the legal team that the *Post* had hired to defend us internationally scoured my emails, which Google—with a recommendation from the Department of Justice—reluctantly let them access after months of pressuring. The strategy was to dissect those emails and, based on the details that the Iranian propaganda machine and judiciary were giving about the charges against me, to anticipate the Iranian government's next moves, which they did with incredible accuracy.

As my brother explained it in an interview, "There are other specific pieces of evidence that we believe they are going to use to support the charges."

The first was a letter I wrote to the Obama transition team in November of 2008, a response to an open call from them for people interested in working for America. Millions of people across the country did the same. Very few received a reply. I was not among those who did.

And another was documents associated with Yegi's application for U.S. permanent residency status. A process open to every single foreign-born spouse of a U.S. citizen anywhere. An incredibly mundane and well-trodden procedural path—even among tens of thousands of Iranians—that had been spun into proof of our cooperation with the enemy.

"What I can say is that those are two of the most significant ones," Ali said. "So I think you can see what kinds of evidence they are basing their entire case on, and that's taken three hundred ten days of my brother's life."

"He'd never do anything malicious to hurt Iran, or the United States," Ali said. "And we want to be as loud and clear to everybody in the world."

Mine was a secret trial with no secrets.

May 26, 2015

BACK IN PRISON

After my first court session I couldn't tell if Kazem was genuinely upset or if it was just more acting. It didn't help that I was back in a tiny interrogation room being talked at through a one-way mirror by him and an angrier guy I never saw.

"Do you know what you've done?" the voice asked. "We wanted to give you a discounted sentence. We have treated you very well. For your crimes you should have been executed months ago."

I had overheard this line about "discounted sentences" before when passing by other detainees in the corridors, who were just as vulnerable, confused, and blindfolded as me. I knew no such thing existed. Sentences were all imaginary until an outside force intervened or the prisoner was offed.

"What crimes?" I asked. "There's nothing more I can do. I've done everything I could. I never did anything wrong and I answered all your questions."

Kazem entered the interrogation chamber.

"We had a deal. What happened? You were supposed to accept your guilt in court," Kazem asked, sounding genuinely hurt.

"You told me to tell the truth, didn't you?" I replied.

"Yes, but then you changed."

"But if I lie in Revolutionary Court, in front of God, won't the punishment be much heavier?"

"You are a very professional man." Kazem smiled. "Your lawyer is not your lawyer. She is just your real lawyer's shadow," he went on. They'd had six months to do any sort of background snooping that they wanted and they were all of a sudden questioning her allegiances, apparently assuming, probably based on her gender, that she would be intimidated into not actually representing me. "It doesn't matter, she will be in here with you very soon. You scored one goal, but we have scored six."

I was pissed and I couldn't keep a lid on it.

"This is Islamic justice? Really? I'm done. Take me back to my cell."

The session ended. The angry guy left. It was just the two of us again and Kazem started walking me back to my cell down the outdoor path, fifty yards I knew well, the ones where we spoke frankly, or appeared to. I was confused, but so was Kazem.

"Why did you do that?" Kazem asked. "He was trying to help you. You are making things worse for yourself."

I was getting sick of these bastards telling me that.

"You bring some guy here to intimidate me and tell me that he's trying to help me? You really think I'm an idiot, don't you?"

"No, J, you are very intelligent. We know that. We have been studying you and your IQ is the highest of any prisoner we have ever had." Any short-term effect of his ego stroking had worn off many months ago, but he stuck to it as a tactic. "Do you think it would be better for you if I come to court next time?" he asked me. That I continued to maintain my innocence was obviously very troubling to him.

"I think I would get angry and tell the judge how you've tortured me and I don't want to upset you," I told him. I was playing with him again.

"Why do you keep defending yourself? I told you to plead guilty and it would finish, and you would go home," he told me as we approached my cell's door. "This is a very real trial."

"So now what?" I asked.

"What do you mean?"

"Will I still leave?"

"Yes. Someday." There was so much gambling going on, but I couldn't really read what was happening around me.

"What am I supposed to do?"

"Plead guilty. It will be better for you." *Do the opposite of whatever they tell you to do,* a voice in my head had been telling me for months.

"How many more sessions will there be?" I was just trying to get any information I could.

"If you accept your guilt the next one will be the last one, but if not, perhaps twenty more sessions. Maybe more." *Okay, now it's obvious he has no idea what he's talking about.* "It is your right to have a trial, but we thought you'd like to go home sooner."

"So if I plead guilty our deal is still on?" I asked.

"Of course," he said, and we were now back at my cell door. The guard opened it and Kazem and I paused at the door.

"Pray tonight. God likes you very much. He will help."

"Okay," I told him, and as he walked away I called, "Come back and see me soon."

"Inshallah," he said over his shoulder. "If I could I would come to see you every day."

Back in the cell I tried to make sense of what had just happened.

"*Vaziat?*" was Mirsani's one-word question whenever I returned to the cell. It meant "situation" or "circumstance."

"Not bad. Not good. Not clear," was my typical answer.

By then I'd realized that the worst they could realistically do to me was put me back in solitary. I feared that to my core, although I knew I had survived it once. But the odds of that were shrinking by the day with all the attention my case was getting.

My treatment had changed so many different times, but now it was clearer than ever that they intended to extract items of value in return for me, and they really couldn't do anything other than continue to hold me and make sure I didn't die.

But that didn't stop intermittent harassments, like the ones that followed all of my court sessions. They honestly thought I was going to plead guilty. They were surprised I hadn't given in yet.

June 8, 2015

SECOND TRIAL DATE

Two weeks passed before my second courtroom appearance was announced. I desperately hoped it would mean the end of the trial. Not only was the self-imposed deadline to complete the nuclear deal, set for July, fast approaching, so was the start of Ramadan.

Having lived and worked in Iran for years and visited other Muslim countries during Islam's holiest month, I knew the court process would slow down considerably and probably come to a complete halt during Ramadan; it doesn't take a genius to realize that when entire societies, in fact whole regions, avoid eating or drinking anything, including water, from sunup to sundown for an entire month, very little will get done during that time. Especially when the ritual cleansing that supposedly comes from this period of deprivation happens to fall in the middle of summer, as it did in 2015. I had been arrested at the height of the previous Ramadan. I couldn't believe they had already stolen a year of my life.

Ahead of that second trial date I requested and was granted—incredibly, I thought—the opportunity to meet with my lawyer, so I was transported to court half an hour earlier than usual and led into Salavati's office.

When I walked in the judge, my lawyer, Leila, and the deputy prosecutor were there.

"We heard you requested a meeting with your lawyer. What was so important that we had to come early today?" Salavati asked.

Amir Ghotbi, the deputy prosecutor who was representing the state against me, sat forward to hear what I wanted to say. I had no doubt that his involvement in my case was the highlight of his professional career, perhaps his whole life. I also knew that the piece of shit had made personal overtures to my wife, asking her at one point why she would marry a spy instead of a nice, educated Iranian (like him).

He even suggested that Yegi should divorce me and start fresh. In the same meeting he'd told Yegi's father not to worry, that I would be released and become a major celebrity, that being arrested was the best thing that ever happened to me. He was a major-league ass-hole.

I asked, "May I have a few minutes alone with my lawyer?"

"What would you need to discuss with her that you can't share with us?" Salavati said.

"Are you serious?" I responded. "You're the judge and he's the guy that's trying to lock me up for life."

"We decided it would be better if we heard what you wanted to talk to her about."

There was absolutely nothing I could respond to that.

"I'm good. We can get back to court."

In the courtroom that day it was more of the same.

I conceded zero guilt and they took me back to prison.

They screwed up with you," Mohsen, a guard who lifted weights and would sometimes be my workout buddy, told me at the gym later that afternoon.

"What do you mean?"

He was the most normal of the guards. In his early thirties, with a muscular build, he claimed to have been a member of Iran's national judo team. I didn't believe him, but I liked him. We joked around a lot. Days that he worked a shift passed a little bit easier.

"They should have put you on trial right in the beginning like they always do. I don't understand why they did it like this. They should have brought you to court when you were still scared and they could get you to say anything. You're not afraid of them anymore."

He was only half right.

"We have been doing everything we could think to do in pursuit of Jason's release," Jay Kennedy, the *Washington Post*'s general counsel, told the *New York Times.* "First, we hoped that the fact that he did nothing wrong would lead to his release. Then, our hope was that continuing discussions between the U.S. government and Iranian officials alongside the nuclear talks would produce positive results. So far, they haven't.

"And so now, we believe, the time has come to bring a very public, adversarial case against Iran, as the *Post* and the family and many others continue to pursue Jason's release through other channels."

One of the court clerks, who proved to be reasonable and even sympathetic, told my lawyer that the IRGC was pushing hard to get Salavati to come on television and denounce me before the trial even ended, a sort of spoiler, as if the outcome weren't clear enough. And he was itching to do it too, but some of his staff convinced him not to. "We told the moron how bad this thing looks already," the clerk said. "Speaking publicly would only make it worse for him.

"Tell Jason he's doing great. Salavati wants to give him a death sentence, but so far Jason hasn't slipped at all and the IRGC and Salavati haven't found any reason that he could. But they are desperately looking for one."

I came close to saying uncle many times during those months, but when I heard this bit of intel any thought of giving in was obliterated by a will to keep going. I had come too far already.

For years I'd stayed in Iran when others had left. Some people looked at me and the handful of dual nationals I called friends as masochists, others just thought we were dumb. It was simple, though, for me. I had to see what was going to happen next. That's the same attitude I took into the closing sessions of the trial. And it didn't hurt that, although it may not have looked like it from the outside world, by then in some ways I was in the driver's seat.

And it showed in their treatment of me.

By the end of my trial Mirsani had kebab being delivered to us once a week, my phone calls and visits were in place, and no one was threatening me anymore. It was simply a waiting game now, but I knew these bastards were experts at that.

July 13, 2015

THIRD TRIAL DATE

The call and response between Tehran and Washington continued. As the nuclear negotiations neared their climax, so did my trial. That kept me in the news.

"We regret that Jason's trial has been closed and his lawyer is barred from discussing the court proceedings," my brother said. "Jason's continued detention is as baseless as it is cruel and unjust. We ask the Iranian judiciary to put an end to the delays in his trial, release Jason, and allow him to reunite with his family."

"Jason Rezaian's unjust detention on espionage and other charges trumped up by Iranian authorities has now, almost inconceivably, stretched into nearly a full year," said Marty Baron, who called on Iranian authorities to "deliver a speedy, fair and impartial judgment in Jason's case—one that could only result in his acquittal, immediate release and a long-overdue reunion with his family. It is long past time to bring an end to the nightmare."

There were other angles to play.

"He is paying the price of the suspicion, the animosity, and the paranoia between the two countries," my mom told reporters between trial dates.

At a press conference marking the historic nuclear deal on July 15, 2015, CBS News' White House correspondent Major Garrett put Obama on the spot.

"As you well know, there are four Americans in Iran—three held on trumped-up charges according to your administration, one, whereabouts unknown. Can you tell the country, sir, why you are content, with all of the fanfare around this deal, to leave the conscience of this nation, the strength of this nation, unaccounted for, in relation to these four Americans?"

Obama was flustered. As much as he had been at any moment during his presidency.

"I've got to give you credit, Major, for how you craft those questions. The notion that I am content, as I celebrate with American citizens languishing in Iranian jails—Major, that's nonsense. And you should know better. I've met with the families of some of those folks. Nobody's content, and our diplomats and our teams are working diligently to try to get them out.

"Now, if the question is why we did not tie the negotiations to their release, think about the logic that that
creates. Suddenly, Iran realizes, you know what, maybe
we can get additional concessions out of the Americans
by holding these individuals. And by the way, if we had
walked away from the nuclear deal, we'd still be pushing just as hard to get these folks out. That's why those
issues are not connected, but we are working every
single day to try to get them out and won't stop until
they're out and rejoined with their families."

Garrett, who covered the entire eight years of the
Obama presidency, said it was one of the president's
most uncomfortable and awkward moments.

None of those moments of public confrontation with
officials—American and Iranian—would ever have
happened if my brother hadn't tirelessly and meticu-
lously pushed every button he could, directly and via
proxies. He racked up hundreds of thousands of miles,
meeting with important people and doing TV and radio
hits at all hours of the day. He simply wouldn't stop.
And he stayed on message throughout: "I appreciate
all of the efforts being done to bring Jason home, but
apparently it's not enough, because he's still in prison."

And it was working.

By the end of my trial plans were being hatched.
American politicians—Democrat and Republican—

expressed their readiness to fly to Tehran to retrieve me. Most of those offers didn't go anywhere, but one of them—from a former U.S. president—was presented to the Iranian diplomats in New York. They were initially excited. They thought it could work. But when they went home with the offer it was rejected by a higher power.

"No, thanks," was the message. "We like what we have in our hand."

August 10, 2015

LAST DAY IN COURT

I n the last session I felt only resignation. This process had started and it had to be completed. That's the rule. If they don't adhere to any of the other ones, finishing the trial is the one thing that always happens. It's part of their efficiency. At least that's what they say.

This session was reserved for the prosecution and defense to read their final positions. In this case that meant those who were trying to lock me up summarized the farcical hearsay and slander of the past year and my lawyer used the law to poke holes in those accusations.

As she reminded Salavati, if there is no crime there

cannot be a conviction, which made it his responsibility to acquit me.

And as Salavati reminded her, "No one will tell me how to do my job."

Sitting across the room from my lawyer Leila that day was *not* Ghotbi, the slimeball prosecutor. He had been replaced by a young, wiry, and uppity mustached female in a black chador.

She talked about the many national secrets I'd sold to America, my close relations with Rouhani and the nuclear negotiating team, and my many attempts to take down the Islamic Republic.

"If these were not acts of espionage to be punished with the most severe sentence possible," she asked rhetorically, "then what were they?" It didn't matter to her that the state hadn't provided anything to back up a single one of these accusations. It was the sort of wild innuendo I'd come to expect, and it was in all likelihood the last time I would be facing my accusers outside the prison walls. I understood: they needed to get their shots in just once more.

But then she said that I had accepted all the prior testimony and interrogation records in my "last defense"—which is what they call the final session of the pre-trial hearings—nine months earlier, and that's when I had to object.

"Excuse me, but I never accepted anything I was forced to say in interrogation. I think that's why we're here now, isn't it?"

She kept going, livid, squawking like some kind of rabid bird, and I just shook my head and laughed incredulously. What else was I going to do?

I didn't try to mask my fears. They were right there with me, front and center, but I had already cried enough. Behind tears of anxiety, once those are all dry, you might find a deep reservoir of indignation. I did. I didn't really know the way to get there before, but I do now. Pondering in the lonely hours what it was that had changed inside of me, I knew it was *that*.

Sure, I'm still optimistic, still hoping for the best. But the force of hypocrites with real power doing whatever they need to in order to maintain their grip, even if that means directing massive resources at a defenseless individual, is a tidal wave that I got trapped under once and survived.

I was exhausted, but my head was up. I wasn't even thinking about what the verdict might be. *Why waste any more time contemplating the obvious outcome of a meaningless exercise? That's a satisfaction I shouldn't give them.*

As the session ended I asked Salavati what was next.

"I'm required by law to deliver a decision within one week."

"Will I be present for that?"

"You will know what my decision is."

"But will I be present?"

"Yes. You will be brought back for the verdict."

"In one week?"

"Yes. Why do you doubt us so much?" Salavati asked.

The driver took me by the arm to escort me back to prison.

"Hopefully it will all be over soon, and you can go back to your life," he said, depositing me into the back of the ambulance, in a rare moment that I accepted as compassion.

After a few knockdowns I was still standing and we were getting into the later rounds. That I was still in the detention center and not the general population, that no verdict had been issued, and that they were still irked by my unwillingness to plead guilty were all small victories that made it possible I might win on points.

12

Waiting Game

When a week passed and then two, it became obvious that a verdict in my case was being held back. Although the trial was over, the drama was not.

My family and the *Post* continued to press my case. Marking something the Iranian government calls Journalists' Day, with no trace of irony, my mom showed up at the Unity Hall, Tehran's most important ceremonial venue.

Government officials gather there every August to pay tribute to the local press corps and maybe choose a few new ones to arrest. Yegi and I had attended the event the year before. She and Mom had heard Zarif would be giving a speech there. As the ministerial limousines began to show up with their tinted windows, Mom positioned herself near the entrance with a sign: "My Son Is a Journalist, Too. #FREEJASON."

She stalked the courtyard throughout the entire event carrying the sign—a protest of one.

It was one of those developments that I probably wouldn't have believed if I was just told about it, but later that day I had my usual visitors at Evin.

"Are you here to wish me a happy Journalists' Day?" I asked Kazem and Borzou when they arrived at my gate. "If so, where's my gift? You look empty-handed."

"Well, J, we wanted to celebrate with you," Borzou started sarcastically, "but your mom and BBC ruined it."

"Oh yeah?"

Kazem pulled out an enlarged computer printout of a BBC web page showing my mom carrying the sign.

"Why is your mother coordinating with the BBC to ruin your life?"

"Maybe you should ask her," I said.

"You know, J, this sort of propaganda against the system is worth two years in prison," Borzou threatened.

"So now you're gonna arrest my seventy-two-year-old mother?"

"No no, we wouldn't do that," Kazem promised. "We'll just add it to your sentence."

Behind the scenes, conversations at the highest levels continued.

In early September the speaker of Iran's parliament, Ali Larijani, who happens to be the brother of both the heads of Iran's judiciary and the laughably titled High Commission on Human Rights, visited the UN for a summit of top MPs from around the world.

My bosses, Foreign Editor Doug Jehl and Executive Editor Marty Baron, trekked to New York, where they would meet with Larijani to press my case. They were able to schedule a meeting with him as well as attend an event with other reporters where they used the opportunity to ask Larijani a question about my case.

They then made their way to Larijani's hotel suite and knocked on the door.

"The speaker will not be seeing you as you asked your question in the press session," an assistant explained to them.

"No, we traveled a great distance to meet with Mr. Larijani privately and we expect that he will honor that commitment," Marty told the functionary, without flinching.

They waited. A few moments later the messenger returned.

"The speaker says he no longer has time for this meeting."

"That is completely unacceptable," Marty replied. "We will wait until he has time."

As they waited by the door, the press secretary of Iran's UN mission, sensing the tension, approached my employers.

"We at the mission are very saddened by Mr. Jason's situation. We have met him many times before"—that was true—"and consider him a good friend." That might have been a stretch.

Marty reached for his jacket's lapel, where there was a #FREEJASON pin. He took it off and held it out for the press secretary.

"If Jason is your friend, I'm sure you will want to wear this in solidarity with him," Marty said.

The man whose job it is to help the Islamic Republic shape its message for U.S. media consumption was at a loss for words. He stared for a moment and then walked away.

Across the Atlantic in Geneva, my brother and David Bowker, a lawyer from the WilmerHale firm, prepared to address the annual meeting of the UN's Human Rights Council. I had been arbitrarily detained for nearly fourteen months at that point.

My brother was there to address the body. Hundreds

of witnesses gathered to hear his testimony. Among them were Iranian diplomats—intelligence officers, really—who tried to intimidate him by photographing him with telephoto lenses, videotaped his speech, and amateurishly tried to record his private conversations with international leaders.

But none of it bothered him.

As he had been since I was arrested, he was undeterred. This was an opportunity to press my case. He had become an expert at that. One of the best advocates for any American ever held in Iran and very possibly in our nation's history.

He was relentless, which wasn't that rare. But he stayed on message and maintained a singular purpose.

That was on a Tuesday in September, the same day of my weekly meeting with my mom and Yegi.

It was a typical visit: I complained that I was going crazy, that nothing was being done to save me, and that if more didn't happen soon I'd be stuck forever. In turn, they assured me—or tried to—that everything that was humanly possible was being done to get me out. Between the three of us we had no concept of what was going on outside of Tehran.

At the end of our hour together I begrudgingly returned to my cell. They were told to stay longer.

A group of IRGC henchmen entered the room and

started lecturing them at a rising decibel level about how destructive the *Washington Post* and my brother's efforts to free me were to my health.

They led my mom out of the room and put her in a car, where she waited.

Back in the meeting room several of them continued to intimidate my wife, but by then she knew the drill as well as me: tell them you'll do what they tell you to do. And then do the opposite.

One of them handed her a prisoner's uniform—the same pink pajamas and chador she'd worn during her seventy-two nights of solitary—and ordered her to put it on.

"Keep stirring up trouble and you will be wearing that for the rest of your life."

After two hours of harassment she joined my mom in the car and they were sent home, where her parents were terrified they had been redevoured by Evin.

All of this was just part of the script of a bad opera gone terribly awry. My captors had forgotten to write the end. Instead they were waiting for the next logical moment to make a splash; I just wasn't sure whether that would mean a heavy sentence, a big release, or some combination of the two.

Looking at the calendar, it was obvious what that opportunity would be: the United Nations General

Assembly. It's the moment that world leaders descend on midtown Manhattan to rub elbows and point fingers at each other, and in Iran's case, when its leaders try to use the mainstream news media to advance their agenda with a wary global public.

Like clockwork Kazem arrived a week before the start of the UNGA and told me I was to be traded for twenty Iranians being held in U.S. prisons. That sounded like a dumb exaggeration even for these guys, but of course I wanted to believe him.

"You will go on President Rouhani's plane with him to New York," Kazem said with certainty. For over a year Rouhani had done everything he could to distance himself from my case, refusing ever to even say my name publicly. He was about as likely to fill his plane with transgender prostitutes—of which Iran has many—to prove the Islamic Republic's progressiveness as he was to give me a lift back to America.

"Yeah, yeah," I told Kazem. "Maybe when pigs fly."

"What?" he asked, smiling as he formed the forbidden and impossible image in his mind.

"It's a proverb."

"What does it mean?"

"It means that what you're saying sounds like bullshit."

"But, J, this is very true. You must be happy. God is helping you."

"We'll see," I told him. "Swear to God?"

"Yes, swear to God this is the plan." He paused. *"Right now."*

Motherfucker.

As the days passed and it became obvious that the judge had no intention of following the Iranian law that mandated him to announce a verdict within a week of the final court session, the media machine working on my behalf pushed itself into an even higher gear.

When a farcical closed-door trial is ongoing and being publicized in an antagonistic foreign capital, it's easier to keep the flame of disgust blazing, but when the news actually stops and there are no more reasons for reporters to show up at the courthouse steps, that fire dwindles to a flicker atop a pile of smoldering embers that needs poking.

Between my brother, the *Washington Post,* and the legal team, they were not going to let the flames die.

It was in those weeks and months that followed the trial that I was elevated into the latest symbol of America's ongoing battle of ideas with Iran.

It's a strange feeling when your life becomes emblematic to both sides in a tug-of-war. It usually happens to people of much greater stature than me. Or after you're dead. The only comparison that seemed to fit was Elián González.

But as 2015 began to wind down and the implementation of the nuclear deal, which would end decades of sanctions and free up billions of dollars of Iranian assets, approached, the quest to bring me home intensified.

The *Washington Post* editorial board wrote about my prolonged detention and Iran's propaganda campaign against me on November 24, 2015, "What could explain this welter of misinformation? Possibly Rezaian is being dangled by the regime as bait for a prisoner exchange. Maybe he is a pawn in a power struggle between the hardline judiciary and the government of President Hassan Rouhani.

"We don't pretend to know. What ought to be clear is that Iran is subjecting an American citizen and respected journalist to extraordinarily cruel and arbitrary treatment—and that it is doing so with impunity."

A few days later that was followed by a statement from the *Post*'s publisher, Fred Ryan, who wrote, "This police state behavior shows that Iran has absolutely no respect for laws, even its own, and exhibits a flagrant disregard for basic human rights."

Fred met with every foreign leader—including Shinzo Abe and Matteo Renzi—that visited the newsroom during my imprisonment, lobbying them to press my case with Iranian officials.

He also made multiple visits to the White House,

where he met with Obama's chief of staff, Denis Mc-
Donough.

"Denis, with all due respect, the president isn't
doing enough to bring our guy home," Fred told Mc-
Donough.

"What more do you want us to do?"

"The president can't even pronounce our guy's name."

"Well, how do you pronounce it, Fred?"

"It's like 'Ryan' with an 'is' in the middle. Ruh-zy-un."

"That helps."

Obama still never got it right.

Zarif and Rouhani, though, knew how to pronounce
it. They were forced to hear my name so often that it's
little wonder that they treated the subject of me with
such open contempt.

On September 16 Khamenei, the supreme leader,
gave what became known as his speech on influence
and infiltration, or what could just as easily be called
"the product of Jason and Kazem's collective imagination in the interrogation room."

I watched in mixed horror and excitement as Iran's
highest power read line after line of bullshit that was
the essence of my interrogation and the "case" that the
IRGC, through its moron agents, built against me.

"An economic and security influence are of course

dangerous and have heavy consequences, but a political and cultural influence is a much larger danger and everyone must be careful," Khamenei said.

I had never done anything that was intended to bring about the downfall of the Islamic Republic. My personal hope was that Iran would someday become an open society. But to my captors this was my biggest crime. And it didn't matter that I wasn't doing anything tangible toward that end. Just my being there to document what was happening was criminal enough.

In his speech Khamenei also warned that the enemy sought influence over Iranian decision makers so that decisions would be made based on the desires of foreign countries.

I couldn't believe it. These lines of the speech were pulled, almost verbatim, from my interrogations. The hard-liners, and their leadership, were even dumber than I thought.

It was a moment of great concern for me. I wasn't sure what would happen next, only that I would become the poster boy for this new propaganda line, which is exactly what happened.

Soon a group of local journalists I had never met, let alone ever heard of, were arrested and accused of being part of my network of influence, on my payroll in fact. Years later some of them are still in prison.

I felt that my life and identity were no longer my own; they had become just caricatures.

The inmates in the political ward of Evin, including another American hostage, Amir Hekmati, laughed when the details of the case against me were aired. When I was finally revealed as Iran's biggest problem of all, they agreed it could not have looked more ridiculous.

From that point forward every time a prison guard or one of the interrogators would enter our cell compound or take me anywhere, I would tell them to not get too close and wash their hands after seeing me because "influence can be contagious."

Kazem was the only one who didn't find it funny. Even he laughed, but it was the chuckle of someone who was smart enough to know that he was the butt of the joke. He and everyone like him had always been a laughingstock and would remain so as long as the Islamic Republic had a pulse. They had been duped the most of all, because they had succeeded even in pulling the wool over their own eyes.

The discourse had taken a predictable turn, and as I often did while I was in prison, I wrote a headline in my head. One perfect for the day. But one that would never see the bold of print.

"Iran Under the Influence."

Within a few hours "the influence" was being tied to rising addiction problems in the society, especially in big cities, where gangs of addicts had inhabited public parks. In the past no one dared touch these subjects publicly or in the media, but now reporting teams were in the parks talking to scared children and even to the addicts themselves.

But I had bigger problems. What the hell did this mean for me? Was this another attempt to raise my perceived value or were they getting ready to do something more astounding and outlandish? Whatever the plan, I felt dangerously close to the epicenter. Iran's political fault lines are hard to predict, and because of the locals' shoddy building methods earthquakes in these parts usually have heavy casualties.

Khamenei spoke about a cabal of people, foreign-based Iranians like me, working in concert to unravel the ideology of the Islamic regime. It didn't take long for me to understand that this speech was a combination of the far-flung accusations of my interrogators mixed with my equally ridiculous responses to their probing questions.

So that's all this was. Create a new boogeyman. And the fact was that they were so weak in their understanding of how the world works that they needed someone else to create the narrative for them.

In fact, there was no cabal. No group of Iranian expats working in tandem with America to defuse Iran's revolutionary ideology. There were only the factors that pointed to a very different path forward: Iran's youth, their educated women, the decline of people's relative prosperity and spending power because of sanctions and corruption, their desire for a better life, and the modern attitude that freedom in 2015 meant the freedom to consume like everyone else.

I immediately internalized the vocabulary of the new speech, which, as it was designed to do, became the newest revolutionary catchphrase. It was called the "influence" or "infiltration" project.

It was around this time that Kazem came to me one afternoon with a request. "Make the media stop talking about you, J."

I had to laugh. "You think even if I wanted to, that I have any power from here?"

"You have the power to stop them. Please tell your mother and wife to tell your brother. And then it will stop."

"Why would I do that?"

"All this attention is very bad for your situation."

"Dear Kazem, my situation is already pretty bad."

13

2015 Comes and Goes

By the end of 2015 I'd grown accustomed to expertly flipping through the lineup of Iranian channels to land on programming relevant to me. Increasingly, after the nuclear deal was signed, that meant live coverage of Obama or Kerry.

This time it was POTUS's end-of-the-year press conference. I had obviously missed the very beginning, because he was already in the middle of a list of his 2015 highlights.

"Around the world, from reaching the deal to prevent Iran from developing a nuclear weapon, to reestablishing diplomatic relations with Cuba, to concluding a landmark trade agreement that will make sure that American workers and American businesses are operating on a level playing field and that we, rather than

China or other countries, are setting the rules for global trade. We have shown what is possible when America leads."

I knew he wasn't going to bring up me and the others still in prison in Iran, but I thought maybe someone would ask. I listened closely anyway, just to feel like a taxpayer for a few minutes.

What has this guy ever done for me? I wondered. *I voted for him. Hell, I even contributed to his campaign,* I told the imaginary guy sitting next to me.

Yeah, yeah, who didn't? he shot back.

I remembered a plan I had hatched in my leanest Rug Jones days, in the wake of Obama's election win. I reached out to contacts in the Tehran bazaar about commissioning a run of several dozen Obama portrait rugs. You could get that done from any high-resolution photo. They send it up to Tabriz in Iran's Azerbaijan province and it takes a couple of months, but what comes back is a phenomenal rendition, usually more real than the snapshot the weaver is working from. The problem was that the price they were quoting me was too high. The sharks wanted about $1,500 apiece, and I needed to make at least thirty of them. I would present one to the White House as a gift and sell the rest. Could have made a fortune on that. But they wouldn't be ready in time for the inauguration. I let it go—because

who knew how long the guy's popularity would last?—
but it was a good idea.

*Yeah, Obama and me. We've come a long way since
then.*

I was suddenly very glad I hadn't gotten those rugs
made. I would have definitely kept at least one. Not
exactly the kind of "evidence" that would have worked
in my favor.

Obama gave himself his end-of-the-year report card
and answered questions about the budget just passed
by Congress, terrorism, and his goals for the upcom-
ing year. And then abruptly—maybe my colleagues in
the room were ready for it, but I wasn't—Obama said,
"Okay, everybody, I got to get to *Star Wars.*"

I paused, shook my head, and let out the same laugh
I had so often as my time in Evin dragged on. The one
that said "Are you fucking kidding me?"

I'd like to see Star Wars *today, too, Obama. Doesn't
matter, though,* I thought. *It'll probably suck anyway.*

I could make a phone call that day. No doubt about
it, the phone calls were a lifeline. For everyone.

Obviously for me it gave me a tiny four-minute
window on the world. Most days there was no news,
but sometimes there was. The calls came in spurts
at first—it was never really clear when they would
happen—but then, around the time my trial started, it

was institutionalized: a four-minute call to Yegi's parents' house late Friday afternoon, and an eight-minute call to my mom's cell phone on Sundays at around the same time.

It was that Sunday call that tethered my mom to Tehran for many months, because everyone knew it would be cut if she wasn't around to receive it for a couple of weeks. Although we didn't feel like it at the time, we had too much firepower for the IRGC to completely have their way with us; they had to make it look like I was being treated fairly, especially when my mom was in the country.

Yegi's family and my mom would all gather together and wait to hear my voice, putting me on speaker. If I was late to call I knew she would panic, inconsolably—certain something terrible had been done to me—until the phone rang.

One Friday afternoon it didn't, because the pay phone system in Evin was out of order and wouldn't be repaired until the following day. I told the sweet older guard who spoke Turkish and was missing two front teeth, the former maintenance guy, that there would be hell to pay.

"Seyed, I realize it's not your fault, but we have to do something. My wife cannot handle the anxiety."

"She knows where you are, and she herself knows how safe it is here," was his response.

I got angry, but not nearly as much as Mirsani did.

"Please tell the shift manager to call my house, otherwise it will be bad for everyone," I tried to warn him. "It will be on international television within hours that I have gone silent." I didn't tell him that I had advised my mom and Yegi to sound the media alarm any time their usual contact with me was cut, even for a day, so that the *Post* and others would be prepared to respond to any irregularities with a story. *Keep my name in the news* became my main mantra, along with *Laugh every day.*

Some days I was more desperate, but I always tried to mask that when talking to Yegi. When she'd visit I could show some vulnerability, maybe even cry a little, because I also knew how to make her laugh again before the end of the visit. But four minutes on the phone, often when she was in mild hysterics due to my calling late, came with a higher degree of difficulty.

I became a cheerleader and a master of slogans.

"Baby, I want you to remember that one day we're going to get out of this mess, and when we do we're going to America. And before long you will be an Ameri-*can,* but one thing you'll never be is an Ameri-*can't.*" I actually said that. And it sorta worked.

Other days the phone calls were tenderer. I knew people were listening, but I didn't care. My wife and I, like many couples, had developed our own way of communicating, making up vocabulary and using words to signify meanings other than their literal ones. Even if they did understand, I didn't care. I was expressing my love for my wife, and I was pretty sure it wouldn't hurt these dipshits to learn a thing or two about romance.

Conversations with my mom were more about giving marching orders, which I realize isn't very nice, but I know my mom and knew that she could handle it.

"Have you been in touch with the *San Francisco Chronicle*? Do they know I'm in here?"

"Honey, the whole world knows you're in there." I only half believed her.

We have momentum, don't let it die.

I was doing whatever I could to contribute creatively to my own freedom campaign. I had the time to think about those sorts of things.

I felt some guilt that so much attention was being paid to me while there were other cases that weren't receiving any. But another part of me, the one that looked out at a horizon of walls, was not as conflicted.

And it was obvious, to me and my family; my lawyers, whom I'd never met; and the U.S. and Iranian government officials doing the negotiating that if I got out the others would, too. If it wasn't for the machine working tirelessly for my release, none of us would go home. Putting it in that light, I didn't feel so bad.

Mom would give me updates on her attempts to contact diplomats in Tehran; France; Italy and Slovenia (her ancestral homelands); and the Holy See (she had Catholic roots and I had spent two years at the Jesuit University of San Francisco). They were literally working any angle.

"You just gotta shake me out of this tree, Mom."

She'd also give me other news. On one call I learned that my Golden State Warriors were going to the NBA Finals for the first time since the season just before my birth. I was initially dejected. *How is it possible that I'm missing this?* It turns out that you miss many milestones when you're locked up.

Just before Christmas 2015 I began to feel helpless. The implementation of the nuclear deal was coming within weeks and once that happened the moments and milestones, the opportunities for a very public exchange, would dry up. At this point, contrary to some previous cases, my detention had become so public that I would not get out of this quietly. That much was clear to everyone.

I told Yegi that I wanted to see a glimpse of 2015 as a free man. It was still possible. I was sure of it. Or thought I was. I ran through all the possibilities over and over again. Leading up to the implementation of the nuclear deal there were so many meetings between Kerry and Zarif; surely they could work out a deal for my release.

For me, as for so many Americans, the idea that Rouhani and Zarif had a mandate to shut down key elements of the Islamic Republic's nuclear program, a program that they'd long claimed was a nonnegotiable right, but that they did *not* have the power to help free falsely imprisoned American citizens was and is unacceptable. A lazy lie that's been repeated so often they have been convinced by a sort of Pavlovian reinforcement that their system's despicable habit of taking hostages is invisible. It's not.

I knew that if the implementation happened and I remained behind bars there might be a brief uproar about my not getting out, but that it would pass quickly enough.

A few days before Christmas on one of those calls home Yegi told me, "Your wish about 2015 isn't going to come true, but it will soon after. In January."

Yegi had a source, someone who worked in the supreme leader's office, who would periodically send messages out of the blue through a third party, providing information that turned out to be surprisingly accurate. Usually it was to warn her of an impending court appearance or some demand Iran was making from the U.S. as part of any deal that would include me.

This time was different, though. Now he was explaining different scenarios of a release that had come close to happening in the previous months, only to fall apart.

According to him the issue of Bob Levinson had been a point of contention throughout the negotiations. Initially Levinson—or a complete accounting of what had happened to him—was to be a part of the deal, but the Rouhani administration decided there was no political value at that time in acknowledging, after eight years, that, yes, Iran had been responsible for his disappearance. They considered Levinson's capture an issue of the Ahmadinejad years and not theirs to solve.

I was to be released with Amir Hekmati, the former marine captured while visiting his grandmother, and Saeed Abedini, a pastor and naturalized U.S. citizen who had been arrested multiple times for converting Muslims to Christianity.

The source told Yegi that I would definitely be out by January, that my release was just a matter of time. Perhaps only hours. Of course there was no way to verify any of it, but her extreme confidence was all that mattered.

I had been dancing on a razor's edge for seventeen months, but never more so than I was then. I thought I could see a light at the end of the tunnel, but also a sealing of my fate. I focused, to the extent that I could, on reading and exercise, withdrawing into routine, knowing that was the safest bet.

My hopes had been up so many times that I refused to get excited, but it did make sense. As much sense as anything had since we were locked up. I needed something to hold on to. The signs were all pointing in that direction. Why else hadn't they issued a verdict in my case after four months?

Christmas started like every other morning. I made a coffee and did a couple of hundreds.

"Sixty-two, get cleaned up," a guard called from the other side of our wall. "You have an extra visit today."

"Are you sure?" I was confused. I had been told not to even request a Christmas visit, that one would be denied. There was a new warden and he didn't believe in making non-Islamic concessions. I hadn't felt like dealing with a rejection so I'd just let it pass.

"Isn't today Christmas Eid?" the guard asked, using the Islamic term for a religious celebration.

"Yes it is."

"Then, yes, I'm sure."

"With my family?" I had learned by then to keep my questions very simple.

"Yes."

"Now?"

"No. Later." He paused. "Says here you can have lunch with them."

Does this mean I'm leaving soon? Or staying longer? Does it even mean anything at all? So many signs to try to interpret and they had all become hard for me to read.

I finished a total of four hundreds, knowing I'd eat well that day. It was the first time in my adult life that I was actively trying to maintain my figure; it was all I had to show for being in prison.

I showered and dressed in my prison pajamas. The new warden had also decided that it was mandatory for us to wear prison clothes to our family visits, apparently to humiliate us as much as possible.

A guard led me through the narrow halls of the ward's offices, a path I had walked hopelessly so many times. But I blocked that from my mind. Today was festive. I could put myself in that space for a few hours. I knew I could.

I entered the room and waited. This was the routine. The only variation being that either Mom and Yegi got there first, or I did. With all the geopolitical complexity of the moment we had been swept up in, this was how basic our family life had become.

When they entered the room it felt like Christmas. They brought gifts and a feast.

I didn't want to eat yet, but I wanted to know what was there. It was dark turkey meat and eggplant, a combination I had never seen, and rice with a bread *tahdig*, the coveted crunchy golden-brown layer on the bottom of a pot of Iranian rice.

They were also able to bring some much-needed fresh clothes. There was a thin zip-up hoodie, a fresh T-shirt, and some clean socks.

There was also a brand-new pair of Adidas cross trainers. When I did the math I realized that, in the just over a year since Yegi was allowed to bring me the first pair of shoes, I had walked the length of the continental United States and then some, and the soles were starting to disintegrate.

And they brought me books.

In the spirit of the day, I was thankful.

Mom and I told Yegi about past Christmases. Where we had been different years, family traditions, and the songs. My dad, I explained, liked "Deck the Halls" so much that besides "God is great," the most often heard sounds from his mouth were probably "Fa la la la la, la la la la."

I tried to stay present with the two most important people in my life, but I slipped into contemplation. It happens when your life is inverted and so much of your daily activity exists alone, in your mind.

I remembered a visit I'd made to the old U.S. embassy in Tehran on a reporting trip. There was a display that showed the hostages around a Christmas tree, and the description talked about the compassion of the Islamic Republic toward the captured spies.

"We treated them so well," was a refrain I had heard for years. "We even let them celebrate Christmas."

"Oh yeah?" I would respond. "What were their other four hundred forty-three days like?" Now I had my answer. Over five hundred of them.

That was what I was most angry about. Not just the guilt that the Iranian regime had pinned on me without giving me an opportunity to defend myself, but being barred from telling my version of events.

A lot of people, in fact, all over the world had decided they could speak on my behalf. I thought to myself that Christmas provided a sliver of a chance for me to speak directly—through my mom—to the outside world; no one else would be talking about our Christmas visit, and if anyone did, it would have been an Iranian official taking credit for giving my family and me this "gift" of time together. I wasn't going to let happen.

For a year and a half every Iranian official who uttered my name lied about me, which is bad enough, but others, including Rouhani, refused to even say it.

All the comments about me made by Iranian officials bothered me tremendously, but it was Zarif's that grated on me most, because he knew better. Since the previous Christmas Zarif had been taking credit for what he called a "humanitarian act" of arranging for my mother to be able to visit me in Evin.

For better or worse, my mom has been a citizen of Iran since the 1970s; she, like any other Iranian mother, had every right to regular visits with me in prison. Zarif knew that it was the only way, short of actually doing something to get me released, to shut people up when they asked about me. If he had taken the time to look he would have seen he was lying through his teeth. But this is an old tactic of chauvinistic authoritarian regimes who have convinced themselves they occupy a moral high ground. Learning to be willingly ignorant of realities obvious to everyone—or at least appearing to be is essential to the longevity of the system and its component parts.

I saw it in Kazem every day, and Salavati on the bench, although he was too dumb to understand it. But Zarif was different. He was the master. He had the full benefit of freedom of movement, a life lived in the United States, an American education, permanent resident status.

He had completely internalized O'Brien's maxim—two plus two is five—despite having had every advantage inconceivable to other Iranians, including a personal relationship with American power. He knew better and he still chose the Islamic Republic. Some people believe there's something admirable in that, I'm just not sure what.

It wasn't his chutzpah or negotiating skills that got the nuclear deal done, it was Zarif and his understanding of America. His attraction to it and revulsion for it. The eternal internal conflict of who you are originally, the essential parts that you're born with, and all that you're able to learn. Classic guilty-conscience stuff. But maybe I'm being too tough on Zarif. He was the key in successfully bridging the divide, even just for a minute. It was something no one else had ever pulled off.

But it was in his public comments about me that he blew his cover, to me at least. He knew better and he chose to stand on the wrong side of history, and not only that, lie about it to a world that also knew better. He lost any credibility he'd had. It was a very clear-cut situation where he and Rouhani could have said, "We don't support the arrest and detention of this innocent journalist." It was that simple.

The fact that he never had to say that, and continues to say I—and others still in Evin—committed any offenses has earned my personal ire. But that doesn't really matter. I knew who he was when he got the job and never expected anything more from him.

On the other side of the world the story was being skewed in another direction entirely. True, it was in my favor.

Friends and colleagues in the journalistic community were quick to come to my defense, and that is forever appreciated, but it still wasn't me talking.

I just wanted to say something. To communicate with those who knew me. It didn't have to be poignant or desperate. I wanted people to know I was hopeful, was still the guy they remembered and could still laugh, even in the face of the absurd gravity of all the forces working against me.

As our family time together began to wane I had an idea.

"Mom, I want you to do something for me. When you leave I want you to send a message right away to the *Post*. There won't be a lot of news today and someone will get back to you right away. You don't need to run this through a committee, just do it." What I meant was she didn't need permission from my brother. He deserved to take the day off.

It was the first moment in many months that I could see a clear opportunity to get my own voice out to the world, add it to the call and response, the "tug-of-war," as my captors themselves called it, over me.

And she got it pretty close to the mark.

"Jason wants all his colleagues at the *Post*, the advertising department, cartoonists, everyone, including the janitors, to know how very much he appreciates their efforts, support, and goodwill. He knows you all are working harder than any other entity to secure his release. And the knowledge of that is what gives him strength every day."

We went over the key details several times until she had internalized it.

"Jason is sending his warmest nondenominational season's greetings to everyone at the *Post* and wishes for a very happy and productive new year."

14

Is This the End?

On the first Monday of 2016, Kazem and Borzou
came to tell me that I was going to be released in a
trade. Eighty-three days since the last time they'd come
to see me—and told me the same thing—had gone by.

"Your friends in America are finally giving us what
we want," Borzou announced.

"What makes this time different?" I asked.

"It's more official now," Borzou said, immediately
making me question the veracity of what he was saying.

What does that even mean? I wondered. I chalked
it up to more harassment. I'd realized months ago that
these clowns were responsible for pulling off my abduction, interrogation, and show trial, but *not* for getting
what they wanted out of the misadventure.

"Why should I believe this when you've lied about everything?" I pointed out in my calmest voice.

"We have not lied at all. Everything we told you was the thruth." Kazem and his colleagues had problems with English words starting in "tr."

"Maybe you think so, but I see otherwise."

"J, we've done everything for you," Borzou started as if giving a report to someone who wasn't involved. "None of our guests have ever had the rights you do."

"I don't have any rights. You've made that very clear."

Like an old married couple my interrogators and I had been having the same fight for a very long time.

"You're an athlete so we let you exercise. Look at you. We helped you get in the best shape of your life." That was true. "Since you have a young wife, we give you legal meetings." That's the term for conjugal visits. That was also true, but only because Yegi was able to convince the judiciary that if we weren't granted them she would expose so many of the other ways our rights were being denied. At its core Iran's legal system actually believes it believes in some form of justice. That's the only way such incredible corruptions of its own laws stand so freely, even against international outrage.

"Okay, Borzou, then let me see your face," I told him. It was an agreement we'd made over a year earlier.

"The day that you leave, J," he said, "that's our deal."

He was right. That was the deal. I decided right then that if and when Borzou showed me his face from behind the surgical mask I would start believing that I was being released. But not until then.

Still, I had other ways of testing their reliability.

It was January 4, 2016. And of course they told me there were some conditions for my release. First I had to ask for a pardon from the supreme leader. I resisted at first, mostly because I wanted to verify with my lawyer, via Yegi, that I wouldn't somehow be giving away a right or acknowledging guilt. They also wanted me to sign away my right to ever sue Iran in any court anywhere in the world, which was a demand I'd expected would come whenever I was actually being set free. I began to formulate a protest, but instead just listened and did all the pushing back in my mind. I had learned by then that it was better to just keep my mouth shut.

"What do you think?" Kazem asked.

"I think you're liars," I told them.

"But what do you feel now that you know you will go home to your family? That everything we said was true?"

I chuckled. Usually that's the best thing to do when you're angry.

After a long pause I said, "I feel like Imam, peace be upon him."

Kazem fancied himself a student of history, Borzou less so. On his arrival to Iran after fifteen years of exile abroad, Ayatollah Khomeini, known locally simply as *Imam*, "guide," or "teacher," was asked by a Western journalist what he felt. "Nothing," was his only reply. I was equating myself to their most revered leader just for the hell of it.

Of course in my head I was bouncing off the walls, racing to figure out what was different about this from every other time they'd promised I'd be released.

We had heard so much bullshit that it was hard to put any stock in predictions or promises, although we desperately wanted to. The signs were there, though.

They gave me their version of events, telling me that in September a deal had been completed, but that America had backed out. Borzou was acting differently now, though, saying that it was a done deal although he would never trust America to follow through on any promise until it was implemented.

"I'll believe it when I see it," I said. So far, in nearly a year and a half of lockup, I hadn't seen anything yet.

They acknowledged that my skepticism was fair.

"I'm like the lying shepherd," Kazem admitted.

"Is that like the boy who cried wolf?" I asked him.

"Yes, I know that one." He smiled as his mind made a linguistic bridge. "It is the same."

"If any of this is true," I told them, "you will come tomorrow to my weekly meeting with my wife and mother. And you'll tell them the same thing."

"Of course we will," Kazem promised. I hadn't seen him in three months. I was more angry with this person than I had been with anyone in my entire life. But whenever he made an appearance I knew there was activity around me. His presence, if nothing else, was always confirmation that I was still relevant.

"Why don't you come around anymore?" I asked him.

"I've been very busy trying to solve your problem," he said. I wanted to punch him so bad.

"Listen," I told him, "if you're lying about this you know you will go to hell, right?"

"Yes," he said. "You should believe me, but I can't force you."

"Let me call my wife every day if I'm really leaving." I was in no position to make demands, but I tried anyway.

"I have to get permission from the judge."

"No you don't. Everyone knows you can do whatever you want with me."

We were walking the path back to my cell. I stopped at the pay phone just over our wall.

"Let's call," I said.

"Wait. I must get the permission."

It was the strangest thing. His guard was starting to come down. He left me there by the phone and was back in less than two minutes.

"The judge agreed. You can call every day, but I have to be present."

"Great."

"Call now, we haven't much time." That was his go-to line.

I dialed Yegi's parents' house. She answered.

"Hi, baby," I said when I heard her voice. This was routine by now, but it wasn't a day that we usually had a call.

"What's wrong? What happened? Why are you calling now?"

"I'm going to be able to call you every day now."

"Why?"

"I just told Kazem that they had to let me and now I can." It sounded ridiculous even saying it, knowing all the imaginary hoops we'd been forced to jump through to do anything for the past year and a half.

"He's there with you now?" she asked.

"Yes."

"Tell him—"

"Please convey my greetings to her," Kazem, who

was standing close to the receiver and listening to as much of our conversation as he could decipher, said.

"He says—"

She interrupted me, "Tell him I curse him every single day for all the lies he tells and the way he has destroyed our lives."

I was about to relay her message but paused.

"You can tell him the next time you see him," I told her, knowing that it would be sooner than she expected. "I gotta go, baby, I'll see you in the morning. I love you." We hung up.

The next morning in the middle of our weekly family meeting there was a knock at the door. It was Kazem and Borzou, as promised. We quickly entered into a heated discussion about where we stood.

"And what about the future and our lives here?"

"That depends on you," Borzou said.

"Oh yeah?" I was frustrated. "On what?"

"How you behave once you're free. A lot of our former guests promise to behave, but once they get out they are influenced by the enemy to act against us. It would be a shame for that to happen with you. We all know how much you love Iran, J."

These fucking guys.

"So we'll be able to come back?" I asked, testing their grasp on reality.

"Why not?" Kazem said. "It's your country."

"She needs to leave with me or I'm not going," I announced.

"That is not in our control. We recommended that, but Judge Salavati did not agree," Kazem lied.

"What are you worried about?" Borzou wondered. "Surely by now you understand the fairness of our system." We were back in the Twilight Zone.

Kazem began a rant directed at Yegi about well-known cases of Iranians—some of them in the political system—who had been imprisoned and became vocal critics of the regime and others who had stayed quiet and came home after some time to live normal lives. He was preemptively, while he still had control of our lives, attempting to coerce us into a future life of silence.

The meeting had gone on for three hours, which was unheard of. They allowed us another twenty minutes alone as a family after the dust settled. My team and I huddled.

"Is this anything different than any of the other times they said I'd be released?" I wanted their view, because I had no perspective.

"Hard to say," my mom admitted.

"Right," Yegi added, "they have lied about everything."

On Friday, January 8, I called home. Yegi was excited. Twenty-five news executives had published an open letter to Secretary of State John Kerry, demanding that everything within the U.S. government's power be done to secure my release. It read:

> *Dear Secretary Kerry:*
>
> *Journalism is not a crime. Yet* Washington Post *journalist Jason Rezaian has been imprisoned by Iran since July 2014 for doing his job. Iran has never offered any evidence that even makes a pretense of justifying this imprisonment. We know you agree that Iran should release Jason and on behalf of our organizations and journalists around the world, we are writing to urge you to maintain your efforts to forge a path to that release.*
>
> *Americans are fortunate to live in a nation that respects the role of reporters and the tenets of journalism. As journalists, we understand how central an informed citizenry is to a well func-*

tioning democracy. The need for information does not stop at the water's edge. Many of our organizations employ journalists who, like Jason, operate in countries, like Iran, that do not always hold a high regard for the free flow of information. We understand the risks involved, and accept them in fulfilling our commitment to provide Americans and audiences worldwide with the information they need to make informed decisions.

At the same time, we depend on The United States and other democratic countries to stand behind the values that Jason represents. Independent journalism is recognized as a fundamental human right. Iran should recognize this, too, and free Jason. The United States has considerable leverage with Iran right now to press that point, and we urge you to continue to do so.

The signers included friends, people who had interviewed me for jobs, producers I'd advised, the heads of some of America's most important news sources, and very influential people who I'd never imagined would know my name, let alone sign a letter in support of my freedom. It was a smart play at the right

moment. America—its media and its leaders—had leverage. They knew as well as I did that if I did not get
out then, I'd probably keep sitting there for a very long
time. I started to believe a little more that my dreams
of freedom might actually be coming true.

The next morning was a scheduled conjugal visit. Yegi
and our lawyer had fought hard to access the many rights
hidden inside Iran's archaic Islamic penal code. One of
these was the right to periodic lawful visits to maintain the sanctity—through regular consummation—of
a marriage. Yegi had discovered that a marriage
is religiously nullified if a man and his wife do not
renew their vows with a carnal act at least every six
weeks.

It had taken many months of my lawyer waiting at
the courthouse for Judge Salavati to see her, but her
persistence had paid off. My lawyer argued that at least
one of the other inmates at Evin Section 2A had the
right to conjugal visits. Salavati retorted that he was
accused of financial crimes, not ones against national
security. But she pressed him on that.

"You and I both know, Your Honor, that Jason
hasn't done anything wrong," she told him. "At least
treat him well so that when he's released this won't be
any worse for the system than it already is."

"Maybe you're right," Salavati said, relenting.

When these visits happened it was a tiny reminder of what life could be. We had four hours to do with what we wanted, in a bizarrely furnished room just opposite where Yegi and I would have our glass window cabin meetings on Tuesdays.

I would be strip-searched at first, but as time went on I was simply patted down. I would bring a towel and a bedsheet, and whatever sugary treats Mirsani and I had at the moment. Yegi has a sweet tooth.

The room had a mattress on the floor, a small bathroom with a shower, and a dorm-style fridge. On the wall there was a sign advising us that there were no surveillance devices in this particular room, which led us to be certain that there were ample surveillance devices in that room.

We didn't really care, though. After the awkward first few minutes—Yegi checking me for signs of physical abuse or neglect and I assuring her I was intact—we were almost able to just be.

It was a hard-won concession that was made begrudgingly. I knew they didn't have to let us have that time, even if the law said they did. But they were trying to placate us the best they could. By then I understood that no matter how badly some people in Iran's state wanted to hang me, they all knew that someday I'd be let go.

My wife and I talked about so many things in those sporadic Saturday rendezvous. About our hopes and fears. And memories. This day was different, though, as possibility, anxiety, and a whiff of freedom loomed. We tried to savor it as the meeting inexplicably started going overtime. It had happened more than once that we were given an extra fifteen minutes, but more often we were gypped, sometimes by as much as an hour. But on that day no one came. Thirty minutes turned into an hour and an hour into two.

"What's happening? Why are they doing this?" Yegi's anxiety, rightfully heightened by a year and a half of attacks on our psyches, refused to let her believe that we had simply been given extra time on that of all possible days. "I think they are going to try to keep me here when they let you go," she remarked. "Jason, if you leave and I don't come with you promise me you'll do everything you can to get me out."

It sounded dramatic, but I knew we had to consider it. She was my wife; of course I would do whatever I had to to be reunited with her. I tried to calm her anyway.

"It won't come to that," I said, trying to reassure us both.

For over a year my brother had held a very solid line to Yegi and my mom that, although he couldn't confirm if anything was actually being done to get me out, any efforts in that direction would include Yegi as well. That was the promise. Our captors, though, always maintained otherwise.

"She is an Iranian charged with crimes, and she must complete the judicial process," was the nauseating answer we kept receiving.

We were battered, but we were still breathing. I wasn't sure how much more we could take.

There was finally a knock on the door. The guard was back.

"I totally forgot about you," he said, flustered. "Don't tell anyone or they will take me off your rotation." He had always been good to us, an accomplice almost, but it was impossible that we would have slipped his mind for two hours.

I set the alarm for 5:25 A.M. Yes, we had a small digital clock. It was one of the holdovers from a past resident.

It was mid-January and President Obama was scheduled to give his final State of the Union. There was no guarantee that Press TV would broadcast it, but given the trend in recent months of showing major Obama speeches, it seemed likely enough. If I missed it I would only get the speech's highlights as decided on by Iran's state censors.

Mirsani had told me the night before that he wanted me to wake him up, too. He never got up before eleven A.M.

Part of me wanted to let him sleep, because I didn't want to have to do a simultaneous translation from English to whatever the mutant hybrid language he and I communicated in should be called.

But he adored Obama, in the way that only someone who has never lived in the United States can unconditionally love an American president. "Friend," he would say, and put his hand on his chest whenever Obama appeared on the screen. "My friend."

It wouldn't be fair to disappoint him so I made enough noise to rustle him out of sleep. After a couple of moments, he turned over, lifting his blindfold to see what was going on.

"Obama starting," I said in our caveman language.

I turned on the TV and my anticipation quickly turned to a state of visceral fear. The news ticker was reporting that the IRGC had captured an American ship in Iranian waters in the Persian Gulf and that the sailors were being held in Iranian territory.

Come on. You've got to be kidding me . . .

All the hope for a swap that I had been led to believe was imminent seemed to suddenly and exponentially diminish. We knew that there were significantly more

Iranians being held in American prisons than there were people like me in Iran. Maybe they were just trying to even out the numbers. Or maybe this was the IRGC trying to throw a monkey wrench into the whole thing. Whatever it was, I was devastated.

And then Obama came on and talked and Mirsani and I listened. The sun wasn't up yet, and we were quiet. "I don't know what he's saying, but his voice is the best," Mirsani said.

I looked at Mirsani and put up two hands, palms up, and looked at him quizzically, as if to ask, "What do you think he's saying?"

It was one of the many routines we had developed with each other. This one we used in lieu of answering a question to which the asker already knew the answer.

"He says: I am good. America is good," he responded.

"Exactly."

We laughed a little.

I waited for some mention of Iran and hoped that he would talk about me, but also for the first time I thought that if there really was about to be a release, it might be better if he didn't. I'll be honest, I didn't know what to think other than that there were some local signs pointing to my release and I wanted one from my commander-in-chief.

After over a year when it seemed as if every one of

his major speeches dealt with Iran in some way, now that the negotiations were finished it was almost not even worth talking about. After all, the deal was not exactly universally loved.

"We built a global coalition, with sanctions and principled diplomacy, to prevent a nuclear-armed Iran. And as we speak, Iran has rolled back its nuclear program, shipped out its uranium stockpile, and the world has avoided another war," Obama said in his only reference to Iran. No mention of me nor the captured sailors.

But the news of the captured Americans dominated Iranian TV coverage throughout the day.

An IRGC general, Hossein Salami—how anyone reporting on Iran can say that name with a straight face, I'll never understand—appeared on TV and talked about the sailors, who, he said, "were crying when they were being captured, but they later felt better after the IRGC forces treated them with kindness."

Afternoon and evening newscasts showed the shell-shocked American soldiers, including a couple of female ones, seated on the floor and being served a basic Persian lunch, supposedly displaying the culture's legendary hospitality. Even to enemy trespassers.

It seemed like a crisis was being averted. I was relieved, but at the same time I was furious.

The nightly news quoted Iran's foreign minister Javad Zarif and Secretary of State John Kerry talking about how their new channel of communication that the nuclear talks opened helped defuse a potentially explosive situation. *They talk every day and meet for many hours every month. Why am I still in prison?*

But I knew there was reason to feel as though there were steps being taken out of view. The nuclear deal was about to be implemented and even the IRGC, parts of which desperately wanted to see the nuclear deal collapse, was acting politely.

"The Americans humbly admitted our might and power, and we freed the sailors after being assured that they had entered the Iranian waters unintentionally and we even returned their weapons," Salami also said.

The incongruity of the thing made me sick: an armed American military vessel illegally—by Washington's own admission—entered Iranian waters, and Iranian authorities were announcing publicly that they had been "assured" it was unintentional in a matter of hours while I sat in prison for a year and a half for doing my job with Iranian state permission.

Kazem arrived later that afternoon for my now-daily phone call.

"Did you see the news today?" He was gloating.

"I did," I responded.

"What's wrong with you?" It was apparent from looking at me that my head had gone to a bad place. "You must be very happy. You are leaving," he told me.

"You sure?" I asked him, unsure.

"Of course I'm sure."

"What about these sailors?"

"It is nothing. They were drunk and having fun in our water, but we showed them it is not a game. This is what happens when you allow women in your military," he said with contempt and amazement.

"So you don't think this changes the trade?" I asked of the same guy who had told me multiple times that the most minor developments—a comment made by my brother on television or one of my own responses to an interrogator—had derailed an imminent release.

"Why would it?" he replied nonchalantly. "Call your wife."

I dialed and when she picked up I could tell she was in good spirits but wound up. I asked her immediately about the sailors and she brushed it off as if it were meaningless.

"Don't worry about it. It's over," she said.

"I know they let them go."

"No, it's over. All of it." She was confident.

During most of my time in Evin the most innocuous encounters had been cause for endless interrogations: a falafel I had eaten on a reporting trip in a contentious region of Iran or the word choice in an email to an old friend. Now, though, I was living in an alternate universe where suddenly even averted international incidents at critical moments in history were of no concern.

I started to believe, just a little bit more, that the nightmare might actually be ending.

15

It's Time to Go

Those nights I wasn't able to sleep. In that sense I had come full circle. All inertia seemed to be moving in the direction of the exit, but there was still so much between me and the door.

I had already made some compromises that I didn't like. After much deliberation, though, in my own mind and with my wife, I decided I would write the letter asking the supreme leader for forgiveness. If they had asked me to do that a year earlier I might have sat down and written a book about how sorry I was for committing all those crimes that didn't exist. But I didn't have time for that now. I had a life to get back to living.

Instead I apologized for any mistakes I might have

made. Kazem looked it over. "Do it again, J. Take out the 'might.'" His English was improving.

I wrote it again, asking for forgiveness for the "mistakes" I'd made. Everyone makes mistakes. I felt dirty but knew I'd get over it.

On Friday, January 15, my in-laws were allowed to come with Yegi and my mom. I hadn't seen them in almost a year. *So this really must be it.*

Kazem entered the visiting room and we had a brief conversation. I told him they all had to be allowed to visit me every day thereafter. He nodded and agreed. He had called my in-laws' house and told Yegi to bring the best set of clothes I had.

On the way back to the cell he pointed to a duffel bag of my clothes sitting in a corner of the office I had passed through dozens of times on the sad march back to my cell. "Those are yours. You're leaving tomorrow," he said, but didn't sound very happy about it.

"I'll believe it when I see Borzou's face," I told him.

He just half-smiled.

Back in the cell Mirsani could sense something was happening. We didn't have enough vocabulary for me to explain all the moving parts. "God willing, good news," I told him in our invented tongue.

My mind raced. I remembered something that

Yadoallah had told me, that when you get out of Evin, he had heard, you can't sleep for the first week. I was getting a head start.

The next day I was up before the sun rose. *Could this really be it?* Mirsani couldn't sleep either. His mother and son were coming to visit that day. We sat down to a very early breakfast, which was his ritual ahead of seeing his family members who would make the five-hundred-mile journey from Jolfa by bus, overnight. Those were the only days he got up before eleven A.M.

When he was ready to leave for that visit around eight A.M., I wondered if I'd ever see him again. We had become extremely close but not touchy. He hugged me, though, for the first time and wished me luck. Based on so many months of negative reinforcement, I told him I'd be there when he got back.

I didn't know how to fill the time so I just started doing hundreds. I worked up a sweat. Around nine thirty the old-man guard came along and told me I was going to court.

I met Kazem by the security checkpoint, blindfolded. He told me to step up and sit in the van. We drove for a minute or two. We were going to meet with a representative from the judiciary, a box-checking bureaucrat from the prosecutor's office for prisoners' affairs who would conduct an exit interview about my treatment, all but confirming that I really was going to be released.

"Is there anything you would like to report?"

"Well, despite being held for a year and a half without due process, being held well over the legally allotted period of time for investigation, and being denied all my basic rights, I don't have any real complaints," I reported.

"So no mistreatment?"

"I have not been physically tortured, but they've done everything else to me," I told him matter-of-factly. And after a pause, "So I'm leaving?"

"This is just a procedure that we do when someone asks for a pardon. I can't make any promises about the outcome, but God willing, something good will happen."

Kazem and Borzou showed up at the cell for one last ball-busting.

"You have to do an exit interview with state television," Borzou announced. "It's part of the contract."

"Let me see the contract," I told him.

"There's no contract," Borzou said. "It's just part of the deal."

"Well you know what else is part of the deal, right?" I replied, motioning to his surgical mask.

His eyes smiled back at me devilishly, as they often did. If this was all an elaborate plan to get me to beg for mercy, only to tighten the screws even further, they were going to very extreme measures.

"Okay, J," he said, and took me around a corner into one of the adjoining interrogation rooms, closing the door behind him. He looked nervous but excited. He pulled the mask off to reveal the face of a middle-aged guy with a big Iranian nose and dimples. His smile reminded me of someone. I thought about it for a second. He was a cross between Jerry Mathers—the Beaver— and Joe Camel.

We'd crossed another imaginary threshold between me and the door. I was getting excited but knew I needed to stay calm, remembering the advice I'd gotten many months earlier from the good prison doctor. "If you stay here for ten years don't get too upset about it, and if you leave tonight don't be too happy."

Take it easy, baba, I could hear my dead dad saying right behind me, just as he did when I was a restless four-year-old.

I wasn't leaving without going back to our cell once more.

I entered our compound knowing it would be to say goodbye to Mirsani if he was there. The door opened and he was waiting outside in our yard. I was wearing my wedding suit. This was really it.

He looked at me, smiling. "*Azad?*"—Free?—he asked.

I nodded.

There had been so many false promises for both of us and this was the first sign that our hope was warranted. I was still nervous and I was happy. But I was also sad, and worried for my friend. He didn't deserve this. No one does.

We hugged again and he started to cry, but he was smiling. I had cried often—sometimes daily for weeks on end—since we met sixteen months earlier, but he only cried when we met and when we said goodbye.

"I'm so happy," he said through his tears. I know he meant it. He was probably the most honest person I've ever met.

They led me away one last time to the infirmary for a final checkup. The doctor took my blood pressure and looked at my ears and throat. I had to sign that I was being freed in physically good shape.

He weighed me. I was down forty-three pounds; this was the lightest I'd been since I came out of solitary. I'd dropped five stress pounds in that last week alone. I knew I'd probably never be that thin again, and I was fine with it.

Next I was led to the office where I'd been processed and had mug shots taken on the first night. It
was where they stashed all of a prisoner's belongings.
My pants and shirt—several sizes too big now—and
a nearly brand-new pair of brown Ecco shoes were
returned to me. I slipped them on and even they felt
baggy on my feet.

The deputy warden who was handling this part of
my release told me it was customary for outgoing detainees to write a letter to the prison staff. Sort of like
signing a guest book, as he explained it, but I took it to
be more like a suggestion box. "If you want the world
to stop criticizing you," I wrote, "give people their
rights. Especially access to a lawyer. It's the law." Even
after all this time I knew that most of these guys still
believed they were doing God's work.

Finally it was back to the room where we would
have our family meetings.

"What are we doing here?" I asked the guard.

"You're waiting for your ride to the airport."

I sat alone for what felt like a very long time. Finally
Kazem arrived.

"Hi, J," he greeted me.

"Hi. What's happening?" I asked.

"Your flight is late. We will go soon," he said.

"What is it? Bad weather?"

"Yes," he lied.

Now that it was all over I thought I might get Kazem to talk, just a little.

"So what was this all about?" I asked.

"What do you mean?"

"I mean why did you do this to me?"

"It's very complicated, J, but you were in a lot of trouble and we saved you."

"No I wasn't."

"Do you know we found a video on your computer of very bad things?"

"Oh yeah? Like what?"

"It was Mr. Rouhani's family. The women of his family in no hijab and they were saying very bad things about our system."

I thought about how ridiculous that sounded for a second and instead of getting in a debate I asked my interrogator, the person most directly responsible for my pain and suffering, "Have you seen it?"

"No," he admitted. "But I heard it is very bad."

I was far past the stage when these sorts of accusations would get me agitated so I just sat and listened.

"You didn't do anything wrong, but when Rouhani learned of this film his people planned to kill you. We had to convict you and keep you here to save you from them. You knew too much."

It was as though he was cramming multiple story lines from different spy movies into one convoluted plot. It was par for the course.

Then Borzou arrived.

"Wow, it's a real goodbye party," I announced.

"We're going to miss you, J," Borzou said in his friendly voice, which always put me on the defensive.

"I bet you will," I said.

"Who will we talk to and learn from?"

"I'm sure you'll find someone new."

"Not like you, though. Someone who will tell us about America."

I thought about their ideological rants and my attempts to answer. I, for one, wouldn't be missing any of it.

By dusk on January 16, 2016, the whole world was reporting that I was free, but really, that's when the final and decisive battle over Yegi and me was being fought.

I was exhausted and a little giddy but scared. Learning to live with constant anxiety is not a skill I would wish on anyone. We had decided that if Yegi was not going to be allowed to leave with me, my mom would stay in Iran until she was able to go. It was the best insurance policy we could come up with, knowing that our hands were tied.

Still, Ali had been telling Yegi for months that any deal to get me out would include her, too. That had been repeated to me so often that I didn't doubt it.

When we left for the airport night had fallen. Borzou walked with us to a van. Everyone else got in and he and I stared at each other for a long minute.

In the back of the van I was instructed to put on my blindfold one last time. We drove the same path to the exit as we had every time I was taken to court. I couldn't believe there would be no hopeless return within several hours. I had been conditioned to see it as an endless loop.

Kazem sat next to me in the back of the van. I tried to pump him for as much information as I could. *Where am I going? Which flight? When will my wife join me? Why can't she come now?*

"We will do everything we can for your wife to leave very soon," he promised. Not only had he lied to me so many times in the previous year and a half, so much of what he said was a full and exact 180 degrees from the truth that I was nervous.

"Jason, do whatever you like when you are free," Kazem advised me. "Write your story and exaggerate it if you have to, but only to make it sell. Whatever you do make sure you get paid." My pious interrogator, who perceived himself as fighting a holy war against infidelity and Western influence, was telling me to "get that money."

"Don't worry. I'll tell my story, but I'll only tell the truth," I said.

"Good. I know. But if you have to give me horns like a monster that's okay, do it."

I still wanted to punch him.

We pulled into Tehran's ancient airport and drove to a building away from the commercial terminals. I had been to Mehrabad so many times over the years, but the only time I had been to any part of the airport not designated for normal passengers was when Josh Fattal and Shane Bauer—two of the American hikers—were released in 2011. The entire Tehran press corps—domestic and foreign—was there that day. I desperately hoped that in my case there wouldn't be a live news "event." As much as I'd hoped for cameras at my court dates to document the injustices I was experiencing months earlier, I was already trying to ease back into a life of anonymity, or at least one in which I controlled how I interacted with the world.

No such luck.

We pulled up to a building that was built for presentations. At the big glass doors there were two bright lights and a camera crew. A mini red-carpet ceremony without the rug just for me.

I got out of the van and the camera focused on me. A state television journalist whom I recognized began asking me questions and I just walked past. In ideologically driven systems propaganda opportunities *always* trump everything else.

I entered a cavernous marble hall lit up with bright fluorescents but freezing cold. There were high ceilings, ornate and uncomfortable chairs, and massive framed portraits of Ayatollahs Khomeini and Khamenei high up on the walls, as there are in most public spaces in the Islamic Republic. I had seen this place on TV before; it was where the Islamic Republic welcomed foreign dignitaries. I had seen video of Putin in this very room.

Kazem and several other agents of the IRGC, all of them in surgical masks, made a makeshift shield around me while the state television crew set up for a shot across the room. In my wedding suit, a lightweight one meant for spring, I was shivering with cold in a room where you could see your breath. Adrenaline coursed through me and I was more alert than I ever remember being before or since. I was getting out.

A couple of minutes later my mom was allowed in the room and we shared a few moments of private conversation. I could see that she was exhausted but staying strong, as she always did, for me.

"Please, Mom, take care of my wife," I pleaded. "Don't leave her behind." I had failed for all these months to take care of Yegi, a very rare promise that I couldn't keep.

"Don't you worry, honey, I'm not going anywhere without her." My mom had put her life on hold for her family so many times before. I hated making her do that again, but this story wouldn't end until we were all reunited outside of Iran. We said goodbye, unsure of when we'd meet again, but I was confident knowing that, with Mom on the ground, no one would suffer any fate worse than what we'd already experienced. That's what I told myself.

Mom walked out and Yegi came in. She was excited, anxious, happy, and scared.

"Do whatever you have to do to get me out of here. Keep fighting," she pleaded.

"Baby, I'm your husband. There's no me without you," I said, trying to calm her.

"I know, I just don't trust these animals," she said.

We sat for a few moments, holding each other tight, not knowing when we'd be together again, and also just freezing cold. But very quickly the television news crew descended on us, lights, camera, microphone. The camera rolled and I just shook my head.

"Don't cry, baby. Don't give them the satisfaction," I told her. She realized it was wise to stop.

"How does it feel to be free, ma'am, and to experience the compassion of our Islamic system?"

"This is not freedom," Yegi said. "You've ruined our lives."

Then they turned to me.

"Mr. Jason, this is your opportunity to tell the world how well you have been treated and how much better we are with prisoners than the United States." I sat silently, becoming agitated, but he kept going. "Please tell the world about your treatment in Iran compared to Guantánamo." *Seriously, this again? Now?* But he wasn't finished. "Yes, and about the treatment of blacks in America by the police."

On their own those were valid points, but I wasn't his fall guy.

"Is this mandatory?" I shouted. "Are we forced to do this?"

And with that, the cameras turned off. I destroyed their scene. The crew withered and receded. But I kept yelling for a few seconds. Yegi calmed me. We were the best team.

A few minutes later the cameraman, a guy in his late twenties with longish hair and a hipster's beard, approached me. I recognized him from a hundred press events I'd covered over the years.

"I'm very sorry," he said. "It's just our job."

I shook his hand, accepting the sentiment but understanding far better than he ever would that he and everyone like him were partners in the survival of the thousand-headed beast of hypocrisy that is the Islamic Republic of Iran. Long before I changed my clothes way back on that first night in Evin I'd known it was going to take many years to root that out.

Yegi was worried about my blood pressure and I promised I would calm down. I had learned to turn my emotions on and off, and right then I powered down. If we weren't going to see each other for a while, I needed to give her a positive parting image. We sat, huddled together, holding each other in that cold room. There was nothing private about the setting, but for a moment it was just us.

"We've made it this far," I told her. "You saved me." Although there were also much bigger forces than us involved, nothing I had ever said was truer. "Soon enough we'll be laughing about the whole thing." My wife just smiled. It was time for her to go, and when she left we hugged for a long time. I kissed her high on the cheek where there was a single tear. "No more crying, okay?" She nodded and left.

I paced that room, freezing. Kazem came to me and said, "J, someone is here to see you. We don't want to let him in, but it is your choice. He is from the embassy of Switzer-land."

"Yes, I would like to see him," I said, understanding that this was the first moment during my entire time in custody that I was being treated as an American.

I sat in a corner and watched a very sturdy man with a cleanly shaven head wearing a great suit glide across the room in my direction, and at first I was skeptical.

"I'm Julio Haas, Swiss ambassador," he told me. It was a piece of information I had no way of verifying, but I took it on faith. "How are you doing?" he asked.

"Cold."

"Where are the others?" he asked. "You are supposed to be three and you are only one."

Who is he talking about? "You mean Amir Hekmati and Saeed Abedini?"

"No, I just came from them. I mean your mother and wife, they are supposed to be here," he told me.

"They have been saying for weeks that they aren't coming. That they will come separately."

"Jason, I have been involved in this negotiation for fourteen months. Your wife is going with you. This has always been the case."

I was relieved, confused, and irate. *Kazem. Messing with me right up until the end.*

"What happens now? Will it all fall apart?" I had been conditioned to assume the worst.

"No, it's too far along," he said, trying to calm me. "It may take some extra minutes, but we will fix this. I have my instructions and I promise you that plane is not leaving without all of you."

"How are they? Saeed and Amir?" It was good to know that I wasn't going through this alone.

"They are together and they are okay. Doing better than you." He rose. "Give me a hug. This is almost over."

The Swiss man whom I'd just met squeezed me and left.

A moment passed and Kazem, who was hovering the whole time, came back to me.

"What did that man say to you?" he asked.

"He told me everything, you liar," I said, trying to maintain calm. "He told me my wife is leaving with me and that has always been part of the deal."

"No, J, your wife was never part of it. This man is no one. He is just trying to make himself a name. Don't let him destroy everything we've done for you."

I wanted to grab Kazem and choke him. That is all I wanted to do. But I didn't. I just laughed. I had nothing else to give. The absurdity had become unbearable.

"Don't talk to me anymore," I told him. "You should just leave."

"Come on, J." He was groveling but still trying to keep the upper hand. "We haven't much time. You must go now."

"Go where? I'm not leaving without my wife and mother," I said.

"J, we explained to you that your wife will leave very soon. We will try to help her," he said feebly.

"Yeah, just like you've been helping me since the beginning? Fuck you," I told him. Cold, exhausted, and hungry, I was running on fumes, but with the knowledge that the Swiss ambassador, my protecting power, wasn't giving the go-ahead for takeoff without my mom and wife, my confidence swelled. "I'm not leaving."

"Let me try to fix it. But this is a big problem," Kazem said.

My interrogator, my only consistent companion on this terrible ride, left the room. It was now well into the night. I paced the length of the cavernous room until I was light-headed with grogginess. I was nervous. I'd been up for forty hours at that point.

I didn't know it, but my mom and Yegi were still in the building. They'd had their cell phones taken from them and been locked in a room nearby after we said our goodbyes.

In Vienna, where the nuclear deal was being implemented around that time, something was off. Iran had agreed to shut down key parts of its nuclear program in exchange for much-needed relief from crushing economic sanctions. That's what the two sides had been negotiating over since well before my arrest.

But the parallel talks over prisoners being held in both countries—a deal that was just as challenging to achieve—wasn't going so smoothly.

The U.S. had agreed to release several Iranians in American jails, most of whom had been arrested on sanctions-related charges. After much deliberation with a Justice Department that didn't believe foreign nationals tried and convicted of breaking actual American laws should be swapped for innocent U.S. nationals being held abroad, they'd relented.

One of those Iranians, though, didn't want to leave prison. He believed he was being held unjustly to begin with and wanted to wait out his sentence to prove a point rather than be pardoned.

Iranian officials couldn't believe it. "Your law can't even *force* someone out of prison?" they wondered. "What good is power if you can't exercise it?"

Complicating matters even further was the fact that all of them were American citizens, too. Dual nationals, just like me and the others being released in Iran. This wasn't a "prisoner swap," it was a release of a bunch of hyphenated Americans, none of whom wanted to return to Iran.

"Just give us one," Iran's envoys pleaded. They needed something to show for this portion of the deal. Releasing me had to have a price associated with it. Absent any returning Iranians, they didn't have that.

The hardline propaganda machine got to work getting ready to spin the story however it could. Another aspect of the historic turning of events was an outstanding American bill. In 1979 the U.S. had seized $400 million that was part of a contract with the deposed shah. Iran had always maintained it was their money and took the U.S. to international court over it. In January 2016 a judgment that would award Iran billions in interest and fines was imminent, so the Obama administration decided to settle the debt to the tune of $1.7 billion.

Knowing full well that the Iranians wouldn't receive any discernible boost from sanctions relief any time soon and wanting to help their negotiating partners in their efforts to bolster their standing at home, the Obama administration had agreed to pay that money on January 16.

Understanding the potentially ugly optics, Obama made the decision: he wanted it to be the beginning of a new era of interaction.

But something wasn't right.

John Kerry's chief of staff, Jon Finer, ran down the tarmac in Vienna to catch a departing Zarif.

"We've got a problem," he told Iran's foreign minister. "Jason's wife is missing."

As ready as Zarif was to put the headache of my imprisonment behind him, the IRGC had another plan: keep my wife as future leverage.

"Figure it out, Javad," Kerry told Zarif in a harried phone call. "Don't let implementation day get screwed up."

Planes sat on runways: empty ones waiting for prisoners to board in the U.S. and Tehran, and one filled with $1.7 billion in cash in Geneva to settle the old debt. The message was clear: if Yegi doesn't get on the plane we don't move forward with any of it.

No Yegi, no money.

The hours dragged on.

I sat and tried to doze, but that didn't work. At one point Kazem came back. He was holding a tablet device.

"J, if you don't leave now we have to take you back

to Evin. It is already after one in the morning. We only have permission to keep prisoners away until midnight. You know how difficult it is to get out of there. If we take you back it might be ten years before you leave." I could see he was as tired as I was. By then he was just going through the motions.

He turned on the tablet and opened a browser. It was Fars News's English site. The main headline was that four Americans had been freed. "Look, J, you're free. Don't disappoint everyone."

"Those guys report fake news all the time." That was true.

"Okay, wait," he said.

He pulled up a similar headline on the *New York Times* website and let me read the opening paragraphs. It was my first interaction with the Internet in a year and a half. I felt clumsy holding the tablet, but I was transfixed as I read the supposed details of our release. All I knew, though, was that I was still in custody.

"I'll believe it when I see it in the *Washington Post*," I told him, feigning disinterest.

He tried to pull it up, but of course the *Post* had been blocked in Iran since the paper started its campaign for my freedom.

I handed it back to him, unimpressed.

"You must leave now, J."

"Nah." I had been around the block enough to know that once it makes the home page of the *Times*—and probably the *Post,* too—it's news. "I'm good."

He left again and I waited. The room was quiet; a few of my guards napped. Others sipped tea from plastic cups and gazed out at the empty runway.

When Kazem came back it was past four in the morning. Walking up to me, he looked dejected and tired.

"Jason, you won," he said.

"What do you mean *I won*?"

"You won. Your wife is going to leave with you," he conceded. "But neither one of you can ever come back to Iran again."

I smiled inside. There was nothing else to say. But he kept talking.

"It's like a chess game. I like to play chess. I'm very competitive, I like to win. But when I lose, I accept it."

He had a piece of paper with him and he told me I had to sign it. For 543 days I had been in the custody of the intelligence wing of the IRGC and they were being forced, in the final hours, to turn me over to the Ministry of Intelligence.

"Jason, you must write that if anything happens to you that we are not responsible once we give you to them," he said solemnly, as if I should be scared.

"Where do I sign?"

Their operation, whatever it was, was over. They knew it. There'd be a new project soon enough, but this was the one they'd be telling their grandkids about one day.

"Where is the Koran I bought for you?" Kazem asked.

"I left it for the next guy," I told him. He was hurt and he deserved it.

They took me through a side door to a ceremonial entryway. Kazem, Siamak, and a few others were there. I looked at all of them one last time. I stared at Kazem for a long time, until we both smiled. Then I did a crazy thing. I hugged him. Yes, it's even possible to develop an attachment to your tormentors, and no, asshole, that's not Stockholm syndrome. It's called being human.

I felt the layers of anger and resentment fading just a little. As my ordeal took on a life of its own, born out of the narrative of convoluted half-truths and conjured fairy tales that became the story of my imprisonment, I had to remind myself so many times of a very important fact: no matter how much this whole thing was directed at me, it really wasn't personal.

I knew that the door was closing on my ever physically being in Iran again. At least as long as these guys ran the place. My wife and I had conducted our lives professionally and socially in a way that was completely transparent. We thought that we could keep the door open. That we could come and go. I'd always wanted to show people that it was possible to live between these two countries. But it's not.

I was loaded into a van, similar to the one that had driven us from our home to prison, but there was no blindfold this time. The driver and the guy who rode shotgun—my new guard, apparently—sat quietly. No one spoke as we drove for just a couple minutes to another part of the airport. They pulled up next to a door and instructed me to go in. It was a sitting room with a half-dozen chairs and two men seated in them: Amir Hekmati and Saeed Abedini. The whole thing began to feel real.

Around dawn, Yegi and my mom were released from their mini detention.

"Thank God Jason has left Iran," the guards lied to them. "Your family's ordeal is over."

They were told to go home and get some rest.

Their cell phones were returned to them and they noticed that they had missed a lot of calls. Dozens of them, in fact, from Ali.

They called him back.

"Where the hell are you two?" he asked.

"We're going home. Jason is free," they reported.

"No he's not. Not yet. Go home and pack your bags and wait for instructions from someone who will mention your favorite dessert. You're leaving just like I always said you would." And he hung up.

They were confused, but not any more than we had been by the last year and a half.

A few minutes later the phone rang.

"Mango sticky rice," the voice said. It was Brett McGurk. Two years earlier Ali had joined Yegi and me on vacation in Thailand, where she discovered that tropical treat, ordering it at every opportunity. She smiled now, knowing things were looking up.

"Wait for a call from the Swiss ambassador," McGurk told her. "He'll get you to your flight."

My mom and Yegi both hurried home and gathered their belongings. My mom had been living in a hotel for months and went to check out. Yegi went home and collected several bags that she'd packed as soon as she was released fifteen months earlier. She said goodbye to her parents, unsure if she'd ever see them again.

By noon we were all reunited at the airport. It was time to go.

16
Freedom Honeymoon

For a year and a half I was separated from the world and made to believe I might never be free again. My life had been turned upside down in every conceivable way, but as I boarded a jet with the internationally accepted symbol of neutrality, the Swiss flag, proudly painted on its exterior, I knew things were starting to look up.

The lone flight attendant, a plump-cheeked woman in her thirties with blond hair, began crying when she saw me. She was Swiss Miss in the flesh and she knew exactly who I was.

"May I hug you?" she asked.

First the Swiss ambassador and now my flight attendant. Turns out that everyone wants to hug an unfairly imprisoned journalist.

At that moment I probably should have realized this story wasn't over yet.

Almost immediately there were signs that my role would still be a starring one, but now, and for this new audience, my character was not actually the villain but the good guy.

As we sat on the plane waiting to take off, there was final jockeying going on somewhere over things we couldn't see but would learn about later, keeping us on the runway for an extra nerve-wracking three hours. We weren't worried anymore, though. If the Islamic Republic wasn't going to let us leave we wouldn't have been allowed on the plane in the first place.

My fellow passengers—my wife, my mom, the U.S. marine Amir Hekmati, and the evangelical pastor Saeed Abedini—and I began to feel the contours of our new lives.

My mom turned on her iPad and connected to the plane's Wi-Fi. She read dozens of emails with messages of congratulations, joy, and love as they poured into her inbox.

From Tehran's Mehrabad airport in the southern and low-lying part of the city, there is a unique vista of the skyline uninterrupted by buildings in the immediate vicinity. The capital looks so ugly, engulfed in smog. I thought about how much I had loved living there and how I missed feeling its pulse while I was imprisoned within the city limits. I knew I wouldn't be returning for a very long time. I felt real relief, but also deep loss.

An official from Iran's Foreign Ministry boarded the plane with a video camera, presumably to provide proof that we were actually all on board. Before stepping off the plane he said in a loud but polite voice, "To the three gentlemen who have been our guests, *don't be tired.*" It's an Iranian colloquialism that is said to someone who either is about to undertake a service or has just completed a task. You would say it to a shopkeeper when entering or to a taxi driver as you got out of their car. It was completely inappropriate in that moment. We all ignored him and he left. With that we were free.

My mom kept sharing news reports about our status on the *Washington Post*'s website and read aloud the emails she was receiving from well-wishers around the world.

There were also clothes for us on board: new white T-shirts, pairs of boxer-briefs, jeans, sweaters, and wool beanies. Three of each, and miraculously it all fit.

It's about a six-hour flight from Tehran to Geneva, but the time evaporated, and by the end of it, having been nourished with veal, chocolate, organic beer, and champagne, I started to feel whole, or at least not hollow, for the first time in many months.

When we were on the runway in Geneva, still on board the Swiss jet, Ambassador Brett McGurk, Obama's special envoy on the Islamic State, entered the plane. "Jason, I'm Brett McGurk. We met once a couple of years ago."

"I remember," I told him matter-of-factly. "You were a character in my interrogations." He was one of dozens of U.S. officials—some I'd met, some I'd interviewed, and many I'd never even heard of—whom I was questioned about for months.

He put his head down, unsure what to say.

"Someday I'll tell you about these last twenty-four hours," he said, struggling to remain professional while holding back his elation at finally seeing the three of us free.

Within days I would begin to learn about the secret negotiations he'd held for many months with Iranian intelligence officers, mostly in Geneva hotel rooms, over our fate. Those talks included the highest direct negotiations between American and IRGC forces in history.

As I stepped off the jet and took my first breaths of free air, crossing the tarmac to get onto a U.S. government plane, I was introduced to several State Department officials.

I was now in U.S. custody. My captors had warned me that if I were ever freed, I would go through a rigorous debriefing and brainwashing by America. I would perhaps even be sent to Guantánamo, they told me.

Looking at these good people, I realized that, while I know my homeland and its security apparatus can be brutal on those who pose a threat—real, perceived, or fabricated—I was now officially out of harm's way.

On that second plane the opulence of the Swiss ministerial perks—the chocolates and champagne—was replaced by a stripped-down American efficiency that I hadn't even realized I'd missed in my years of living abroad. I sat in the back row, single seat on either side, of a ten-row plane. Up front were Amir and Saeed.

They were being interviewed by federal agents about Bob Levinson, an FBI man who had gone missing in Iran in 2007. "Tell them I have no idea about the whereabouts or conditions of anyone else in Iran. I was in isolation the whole time," I announced, but no one was paying attention to me. Everyone was too happy about our freedom to worry about what I had to say.

We were offered food. In accordance with Iranian social protocol, I said that I wasn't hungry just then. I had just gone through an incredible year-and-a-half-long ordeal that culminated in a couple of historically tense days, and I'd been deprived of so much for so long. I could have eaten anything they put in front of me, but declining an initial invitation had become habitual.

And in typical American fashion there was no second offer. *Welcome home,* I thought.

When we landed in Germany I was quickly and unceremoniously separated from my mom and Yegi, and directed aboard a long yellow school bus, where several soldiers in camouflage sat quietly. It felt as though Amir, Saeed, and I were going on a field trip.

A handsome black man, bald and with a goatee, sat down with each of us and gave his pitch. He was the lead SERE (survival, evasion, rescue, escape) psychologist, we were told. We were on our way to Landstuhl Regional Medical Center. We'd be getting the best post-captivity care in the world. Medical and psychological.

And of course, it was all "completely optional."

It was a snowy Sunday evening in January, so it was hard to know what time it actually was. I'd been awake since dawn the day before and had only slept a few hours per night for the previous ten days. I was running on the pure adrenaline of hope fulfilled.

As the bus approached a compound I could see a large gate, like a toll plaza. It was lit up, but in an exaggerated way. I began to recognize all the accessories of live television reporting: the trucks with their satellites and the lights that can turn the gloomiest winter night into a giant outdoor tanning booth.

At that moment it didn't occur to me that it was all for us. *Americans get released from foreign prisons as pieces in historic and high-profile geopolitical deals and sent to military hospitals all the time, right?*

We passed through the gates, entered the sprawling grounds of the facility, and pulled up to a standard hospital entrance. As we stepped down from the bus there were three wheelchairs waiting. It took a second to register that they were for us, too.

"We got some bad intel," Dr. Carl, the psychologist, admitted a little sheepishly.

Amir, Saeed, and I paused for a moment to consider the chairs, and then walked ourselves into the hospital. We had all just spent significant time in prison, and although we each looked much thinner and older than we did in the pictures that were used in news reports throughout our imprisonments, we all felt pretty good.

Staff members took us up to the third floor of the hospital, to a section where some kind of security clearance was required to enter. In every other way, though, it looked like a section of a typical American hospital.

The only difference was that we were the only patients there. Each of us was given our own room, far down the hall from one another.

Soon a doctor came and gave me the most cursory checkup. "We have reason to believe that at least one of you *may* have been exposed to tuberculosis while in captivity, so we're going to go ahead and give you a routine TB test. Once we do that, though, we're gonna have to keep you here until the results come back."

"Whatever," I told him, just thankful to be dealing with someone who spoke English the way I do.

A nurse came and swabbed an oversized Q-tip under my arm.

"How long is it going to take to get the results?" I asked.

"Probably three to five days," the doctor responded.

That sounded like total bullshit to me, but I wasn't ready to rock any boats yet.

After a high-stress, high-sweat forty-eight hours, I took a shower. A nurse had given me a kit of American grooming products: Head and Shoulders shampoo, Right Guard deodorant, a Mach3 razor, and some Barbasol women's shaving cream, because they were out of the men's.

The water didn't get that hot, but it had good pressure.

I took my time cleaning up. I could do whatever I wanted now, but I realized I was still in a sort of captivity. I could do whatever I wanted *in that room*.

I wanted some food. I poked my head into the hallway and asked a nurse what was possible that late on a Sunday night. She brought a menu. Hospital food. I ordered a grilled cheese—probably served with a tiny tub of pudding or applesauce, I guessed—and chicken noodle soup.

I lay down on the adjustable bed. It wasn't very comfortable and I was wound up. "Would it be possible to get something to help me sleep tonight?" I asked the nurse.

"I'll ask the doctor when his shift starts," she responded dutifully. Chain of command.

Next to the bed was a bulky TV coming out of the wall on a moveable metal arm. I adjusted it to face the bed and turned it on. There was static, but it was watchable, and best of all, after over a year of only Iranian state television, everything on the screen was in American English.

It was Fox News and the reporter looked very cold. I turned up the volume. She was live on location in Germany, where some Americans had just arrived and were receiving much-needed medical care. And then the pictures of my fellow ex-cons and me appeared on the screen.

"Little is known about their conditions, but they are free after their long ordeals, getting the medical attention they so desperately need."

I had to laugh a little. I wasn't completely free yet and I wasn't actually getting much medical attention. But I was riveted, and I kept watching. How could I stop?

The nurse finally did get me a sleeping pill, and when it kicked in I was out.

Eventually, I was allowed to see Marty and Doug, my bosses from the *Post*. We sat in a small meeting room in some other wing of the hospital. It was so soothing to know that I was being supported by my employers and that they didn't want anything from me.

Later I was able to see my brother, Ali. It was my first contact with him since a phone call the day before I was arrested. He had done so much to get me out, some of which, I'm certain, he'll never share.

But thankfully, true to all I know about him, he didn't feel the need to cry or hug me. I threw an arm around him anyway.

"Hey, Jason," he said when he saw me, and it was

right then that I realized that not everything in the world had changed.

By day three Yegi and I were sleeping in the same bed for the first time in a year and a half.

That felt like a victory.

All the while the reporters waited outside the base's gates no more than a couple of hundred yards away from me. I started to feel bad for them in that cold, but really they seemed so far away.

Soon the talk turned to how to deal with them. "Just tell them I'm not coming outside for now," I told Ali, thinking that would magically get them to go home.

"Jason, they're not going to leave until you leave," was the unwanted response I kept getting.

On our second day there Amir, the marine, had gone out to address the reporters with his relatives and congressman. When I saw the footage of that I knew I was nowhere close to ready to interact with the media. I recognized the look in his eyes and the struggle not just to find words, but to trust that they were the right ones. Maybe that's just projection, but I put myself in his shoes and knew that I was not going to even consider going on TV for a while.

"Why don't we just make a statement?" I asked at some point.

That was apparently what everyone on my team wanted to hear. That group consisted of my brother, Doug Jehl, and Bob Kimmitt, the lawyer the *Washington Post* had hired to represent my family and me in international courts as they fought to hold the Islamic Republic accountable for its illegal—by international standards and its own laws—detention of me.

"I'll write something up and you guys can edit it?" I suggested.

It took me a couple of hours to actually sit down and write, but I did it and delivered it to my editor, just like old times. In it I thanked the world for its support, promised to tell my story, but asked for time to wrap my head around what had just happened to me. Us.

They liked it, but I felt like something was missing. I needed to personalize it a little.

"Let's add that I'm looking forward to seeing the new *Star Wars* movie and watching some Golden State Warriors games." Doug nodded that that was a good idea. An hour later it was online.

I should have said I wanted so much more, because I probably would have been given all of it.

Before the end of the day I had been invited to have a private screening of *Star Wars* and to watch a Warriors game, courtside. The lesson was clear: go to foreign prison, get out, and then get cool stuff.

How nice it would have been if great invitations and gifts were all I got . . .

At that point, though, the freedom honeymoon was still in its earliest moments. There were no headaches or tough decisions. No one had expectations. Or maybe they did, but I just couldn't see them yet. So much had been taken from me, and I was starting to get some of it back. And that felt good.

I wanted some jeans. None of my clothes Yegi had so lovingly packed for the day I'd finally be free came even close to fitting.

Dr. Carl and an employee of the hospital took me to shop at the local PX.

A PX, for civilians, is a place where military personnel abroad can go to be American. I imagine them in the most far-flung corners of the world, wherever American soldiers are based. Picture your local mall, with all the same shops, a food court with a Pizza Hut and a Popeyes, a multiplex, and plenty of parking, because it's in the middle of nowhere.

We walked in and for the first time I felt the warmth of America: so many fellow citizens from so many different backgrounds spending their hard-earned paychecks on stuff they didn't need and food that would kill them. I knew this place. It was what I'd moved away from seven years earlier. It was home.

"Let's just get in and out," I told Dr. Carl, invoking my longtime rule for going to the mall. I needed some Levi's, maybe two pairs, and a belt. I let myself be guided to where I instinctively knew men's jeans must live.

"Jeff Bezos would like to come pick you and your family up with his jet if that would be all right with you," Bob Kimmitt told me when I returned from my shopping expedition. After serving in the army, earning a Purple Heart and other military honors, Kimmitt had became U.S. ambassador to Germany, and later Deputy Secretary of the Treasury in George W. Bush's cabinet. At that point I just knew him as my lawyer, Bob. I had been out of prison and back with my family for four days.

"We want that, right?" I asked my team.

"Yeah, Jason," my smart-ass brother replied. "It beats Lufthansa."

The plan was that he would arrive, we'd all have a nice dinner together, and we'd fly out the next morning.

The only problem was that he wasn't automatically cleared to enter the base—though it was assumed that

that could be worked out—and Yegi couldn't leave the base, because she had no passport.

Since we couldn't go off base, Jeff Bezos graciously came to us. And we ate takeout schnitzel and drank beer at the base's Fisher House, part of a network of facilities where the families of servicemen and women receiving medical care can stay free of charge.

From the moment he arrived on the base Jeff was completely engaged with us, immersing himself in the situation, even attending a session with the psychological team, where, for nearly an hour, we discussed many of the symptoms of post-captivity.

At dinner that night Yegi and I recounted some of the ordeals we'd endured over the past year and a half and we laughed. Jeff Bezos has an incredible, booming laugh.

We went to bed early that night. The next morning he came in a shuttle bus to take us to a landing strip half an hour away. Not even Jeff Bezos could land his plane in Landstuhl. I will always be humbled by all the effort he went through to get me home, and the amount of time he personally spent to make sure I actually got there.

Friends and colleagues had asked for my opinion when Bezos bought the *Washington Post* in 2013. I didn't know enough about him then to have one. Now here we were sitting across from each other on his jet, which was festooned with streamers and #freejason posters and stocked with guacamole and burritos— "Freedom Burritos," we joked—made on board, because he had heard I liked them. They were delicious. You can have just about anything on a private jet with a little advance notice, I learned.

We were flying first from an airstrip in central Germany to Bangor, Maine, the frozen northeastern-most international airport in the continental United States. Looking down from thirty thousand feet I could feel the cold below, as we passed above icy rivers and snowcapped pine forests.

Bangor was chosen as an easy entry point that wouldn't have much traffic that time of year. As it turned out we were the only passengers that late Friday morning. After landing we were met by a massive Homeland Security officer with red hair and a bushy beard to match. If he'd removed his uniform he could easily have passed for a Hells Angel.

"Welcome to Bangor, Jason. We're very happy to see you here," he said in the local accent, giving me a big bear hug of a welcome. "You screw with one of us, you screw with all of us."

He led us toward the arrival hall to officially process our entry into America, but for me that happened when I knelt down and kissed the frozen runway. My brother had made me do it, and he took a picture.

Once we were inside, Bob and two of the officers took a nervous and passport-less Yegi to an immigration window. The rest of us—my mom, my brother, my boss, Jeff Bezos, and I—waited.

It didn't take long. After less than three minutes Yegi returned with tears in her eyes.

"What's wrong?" I asked her, worried that something was amiss.

"I'm so happy about how well my new country is already treating me, and so ashamed for how badly my own country did," she said through sobs.

There was nothing more to say.

In an empty and frozen New England airport arrival hall hearts melted. Jeff Bezos and my family had a moment. We were all in this together. The big redheaded Homeland Security guy gulped and his eyes moistened.

We'd been on the ground for less than half an hour. It was my first time in Maine. I had now visited my forty-sixth U.S. state.

Although Yegi didn't have any documents proving her identity, our new friends at Maine's border control issued her a handwritten paper I-94 visa with her status listed as "humanitarian parole." Usually that slip of paper without a photo would be affixed inside a passport. For the time being, though, that was her only acceptable form of ID.

Given all the planes I've been on, "turbulent" is the best way I can describe my time as a freed man; it's been a mixed bag of ups and downs that often feels as unfamiliar to me as my time in prison did. The difference being that after such dark times I am filled with irrational hope about the future.

But that's probably because I don't watch cable news.

We met President Obama, twice; hung out with Bette Midler—a wonderful lady; and were fitted for expensive party clothes that we didn't have to pay for. Which is all great, but interspersed with all those encounters with high-ranking government officials, celebrities, and billionaires have been countless hours of bureaucratic hoop jumping, dozens of flight segments, and so many of the other aspects of a life in America that none of us love.

In April 2016 Yegi and I celebrated an uncommon milestone: we had finally spent more days of our marriage together in freedom than we had separated by Evin's high brick walls. That happened on the eve of our third wedding anniversary.

We have to remind ourselves that we are still relative newlyweds, trying to navigate a life in a new environment with very different circumstances than the ones we knew when we were detained on July 22, 2014.

We spent our anniversary as we did so many other days during the first months of our marriage: filling out forms. Although we had completed the entire immigration process for Yegi to become a U.S. permanent resident before we were arrested, we had to start from scratch when we arrived. All of those completed and approved applications had expired.

Anyone who has done any time as a journalist in an authoritarian Middle Eastern country like Iran or as a recent immigrant to the United States, or the spouse of one, will tell you that completing paperwork is one of the defining tasks of both pursuits.

I'm still staring at a stack of papers. Many of them I can just shred, but others require my attention. Some even call for immediate action. *If I don't respond what's the worst they can do to me?* I wonder.

I haven't done my taxes for the last three years. Apparently they can't be filed without the taxpayer's actual signature. I've got several bills in collection that never got paid, because I was in prison. These are things, I've learned, that don't just magically fix themselves.

And I've already been summoned for jury duty. Like most people, though, I got out of it.

My mornings start early. It's a good time for thinking. Plotting, as everyone else sleeps. Most of those thoughts, unless they get typed, will fade into nothing. My mind—my memory and focus—is still mush.

Mornings feel good, probably because the solitude doesn't feel forced anymore and there are naturally far fewer distractions. That feeling of focus that comes from having so few options is probably the only thing I miss about prison. It's the one time of day when I don't feel I'm a total mess. Everything begins to unravel with the rising sun.

Yegi and I misplace things. We get agitated about it. I'm pretty sure it's temporary. She's a perfectionist and gets angry at us. We've aged. Neither of us trusts anyone anymore. But we trust each other. We've been through too much together not to.

I become confused in crowded places. I don't like talking on the phone. I get recognized when I least expect it, by people who I still can't believe know who I am.

It's never as comfortable as I make it look.

Soon after my return to California I went to the local DMV to renew my expired driver's license.

As I waited in line, a woman in her fifties and wearing sunglasses inside sidled up beside me.

"Are you Jason?" she asked in a hushed tone.

"Yeah," I replied, equally muted.

"I followed your case. You've been through a lot," she told me.

"This is true," I agreed, and nodded.

"Everyone, this is Jason Rezaian, the *Washington Post* reporter who was in prison in Iran and was just released," she called out to the crowd, as if that were necessary *or* appropriate.

Right there, at eight A.M. on a Tuesday, I got a standing ovation in a place where that sort of thing isn't supposed to happen. I still had to wait in line just like everyone else, though.

"Jason, it is so good to see you," strangers tell me as they ignore all our unwritten rules about appropriate personal space and hug me.

"Well, thanks," I tell them. "It's really good to be *seen*." Everyone loves that. The truth is I'm really not sure what else to say.

The low-level anxiety that rang constantly in my head during my time in captivity has not gone away completely, but it's dissipated, returning with headaches in places I've never had them before whenever I try to do too much. There is too much to do, although very little of it actually gets done.

After a year and a half during which nearly all of my experience of transportation consisted of literally walking in circles, I am now behind the wheel again.

Thankfully I'm experiencing less car sickness than I did during my first few weeks out in the world. That said, I've already been in an accident. My oldest friends will tell you that was to be expected, prison or not.

Any fear of flying that I used to have has been replaced with wonder: I am actually free to go places again. It helped that for a brief time—way too short, in fact—when all Yegi had as ID was that I-94 visa and couldn't board commercial flights, we continued to be ferried around the U.S. by private jet. Thanks again, Jeff Bezos.

With the help of our congressman's office and the Department of Homeland Security we've solved that problem, and we now get to ride coach, invariably sitting in middle rows on opposite sides of commercial airplanes, just like every other American who books tickets last minute and doesn't have any frequent-flier status.

Months after my release I still haven't shaken all the paranoia. I get tired early and can't stay asleep. When I was locked up I dreamed of being free. Now that I am out the only dreams that I have are nightmares that I'm back in prison.

Yegi often shakes me awake. "Jason. Are you okay? You were screaming again."

I think she's worried that a part of me is broken forever, but that's not something I'm willing to concede.

"I just got out of prison," I remind both of us. "We're going to get through this," I promise, remembering that I've kept all the promises I've made to my wife so far, unsure of how I plan to keep this one.

It was two P.M. in Washington, DC, on the afternoon of the 2016 White House Correspondents' Dinner. We had been back in America for about three months. In a few hours I would be at "Nerd Prom," sharing a podium with President Obama. A year ago he'd used the same stage to talk about press freedom and spoke of his commitment to getting me out of Iran. Now I was free.

In our hotel room on Fifteenth and L, Yegi was in full hair and makeup mode. My wife wakes up gorgeous, but when she gets going with the products she becomes otherworldly.

"Jason, we need lunch," she told me in a tone that I knew meant I was to fetch us lunch.

"What do you feel like?" I asked her.

She was so confident of her order that I hadn't even finished asking.

"Five Guys," she replied with the smile of a little girl who knows she's doing something wrong and is trying to preempt discipline with cuteness. I can never say no to that smile. And on this particular occasion, why would I have? I wanted a Five Guys, too.

I exited the hotel onto Fifteenth Street, where there was a cacophony of sound. I'd unwittingly booked us at a place on the same block as the *Washington Post*'s former newsroom—as it was being demolished.

Walking past it slowly, taking in the scene of crumbling cement and the clanging of proud and resilient steel girders that refused to fall, a voice called from behind me.

"Hey, brother."

I decided to stop, although I told myself I should keep walking.

I turned around and a tall black man about my age was standing before me, not disheveled but not doing very well either.

"I need a little help," he started his pitch. "I don't use, and I work. I ain't looking for handouts, just some help."

At this point one of two things usually happens: you either walk away or dig a hand into your pocket.

With only a moment's contemplation I chose a third path, and almost immediately felt like an asshole.

"I can't help right now," I told him, "I just got out of prison."

"Oh yeah?" He leaned back to size me up. "You, too? How long they did you for?"

"Well, about a year and a half," I responded.

"Did me for eight years. Just got out Thursday," he responded knowingly.

All of a sudden I didn't feel so cool.

"What they got you on?"

"Well, that's sort of complicated."

"Oh yeah, why's that?"

"I was in a foreign prison."

"Oh, so you was in the service?" He was genuinely interested.

"No, I worked for them," I said, pointing over to the carcass of the old *Post* building.

At that point he took a long deep look at me and pondered.

"Hold up. You that guy. You that guy from the news. You Free Jason." He put an arm around me and shook my hand. "Yo, Malik, Dawood! Get over here. It's Free Jason." His two friends made their way over.

"*Salaam alaikum,*" I greeted them, assuming from their names they must be brothers from the Nation of Islam.

"*Wa alaikum assalaam,*" one of them responded. "How you doing? Man, we were praying for you." I'm not sure if he was Malik or Dawood.

"That means a lot to me." And it did, too.

"Man, what you been through, that's rough. We were following your story from inside. Everyone was with you." I found that a little hard to believe, but by this point very little surprised me anymore.

"Listen, guys, it's been a real pleasure, but I need to get going."

"No problem, Jason. Hey if you need anything, man, we got you. God bless, brother."

Only in America, I thought, and kept walking to get those burgers.

Epilogue

Fortunately our lives began to normalize. Some-what. By the summer of 2016 I could sleep through most nights without any pills and the instances of my name being in the news started to decrease.

Yegi was still virtually stateless but had been issued what is called a travel document, essentially a passport for refugees and others who don't enjoy diplomatic coverage by their native government.

Once we had that in hand we made plans to meet her parents for a much-needed reunion after our ordeal.

After ten days together we reluctantly parted, they back to Tehran, and Yegi and I to Washington, DC.

It was August 3. When I turned on my phone after landing at Dulles I had several texts and emails asking me for comment or to appear on that evening's news.

Obama Doctrine. That's a large part of why I'm so torn on how my detention was handled. Pragmatism trumped the vitriolic chest pounding that passes as patriotism.

I got out, but maybe I would have been freed sooner if the administration accepted the challenges posed by Republican hawks that freeing Americans should be a prerequisite for any negotiations. They didn't, and I spent eighteen months in the joint, the plaything of some of the nastiest authoritarian ideologues to roam the earth in many decades.

But a more obstinate approach might have just as easily backfired and I could have remained stuck in Iran until now, just as the other Americans who remain there to this day.

Obama officials admitted as much when I met with them on my first visit back to Washington, which coincided with my fortieth birthday and the Iranian New Year. The latter is an occasion that had become an opportunity for Obama to extend a virtual olive branch to the people of Iran.

Two things became very clear when I met Obama. The first and most important right then was that we were a small group of people just having a personal moment.

Obama, Secretary of State John Kerry, and National Security Advisor Susan Rice were all in the room, and

As we deplaned I checked the headlines and there it was: Obama paid 1.7 billion dollars to Iran to release U.S. hostages. That day the *Wall Street Journal* had run a front-page story that the money that had been given to Iran on the day of my release had been cash. This I knew, as did several other national security reporters who had been working the Iran story. As Ben Rhodes told me the month before, when I interviewed him for this book, "It's Iran. You can't write a check or something."

Everyone knew about the money—Obama had even announced it as part of the details of the nuclear deal's implementation. But that it was actual cash—paper bills—apparently made it newsworthy again. Although the Obama team wasn't completely up front about the cash, they weren't actively trying to hide it, either.

The story took on a life of its own. It was a textbook spin job, weaving something out of virtually nothing. One of the many nonissues that became an essential part of the issueless presidential race of 2016.

The same sorts of idiots who had blamed my arrest on Anthony Bourdain now wanted to know how it felt to know my value. I became the "One-Point-Seven-Billion-Dollar Man."

The money wasn't a ransom, but it did offer insight into the Obama administration's approach to solving problems.

The story of that money is pretty clear: a court in The Hague was going to rule in favor of Iran in a long-standing legal dispute between Tehran and Washington over Iranian money paid to the U.S. for an arms deal that was never completed. Once the revolution ended the U.S. held the money.

The administration understood realities that their Iranian counterparts had either missed or not properly accounted for, most importantly that Iran wouldn't receive much financial benefit from the lifting of sanctions in the immediate future.

They also knew that the U.S. was going to have to pay a vastly larger amount to settle our debt to Iran when the court ruled against the U.S., which it was set to do.

On the issue of American and Iranian prisoners being freed, we were technically a separate issue: people for people. Me, along with the other Americans released with me, were exchanged for several Iranians imprisoned in the U.S. on sanctions violations charges. It turned out that all of them were also U.S. citizens and none of them wanted to return to Iran.

The Iranians had egg on their faces. The propaganda opportunity that the prisoner swap promised to be fell apart. That was Iran's fault, probably, but frankly their American negotiating counterparts did not want to see them fail.

Obama saw it as the start of a new day with Iran. An opportunity to build a relationship—not an alliance—toward a working peace in a part of the world that hasn't known it in generations.

But that was irrelevant to opponents of the deal in Congress and the alt-right media. It was as though I were living in the mirror image of what I had just escaped.

The Iran deal, for better or worse, will go down as Obama's signature foreign policy maneuver. Deft diplomacy or capitulation? Letting a rogue regime run wild or a calculated move that will defuse the threat of an old ideological enemy? History will decide, but either way I'm stuck in the middle of this thing, the person most identifiably attached to the story of Obama's most important foreign policy legacy.

Now that the nuclear deal and Washington's period of actual engagement with Iran is a relic of the past, it's easier to judge. As far as I'm concerned, the whole thing was a wash: it got me arrested and it got me released.

I have always believed in the principles that were at the heart of Obama's openings with Cuba and Iran—the notion that talking with other nations, even if they're our adversaries is always preferable. The

all of them, as well as Yegi and I, got a little emotional. They expressed a sense of personal responsibility that frankly I didn't expect.

Obama made it clear why: the effort exerted by my brother and the *Washington Post* set the tone for everything that was done on my behalf.

"Your brother made sure we never forgot you," the president told me. "He kept the pressure on us, so don't give him a hard time. At least not for a few months."

John Kerry told me I had a lot to be thankful for, including my beautiful wife, a chorus I hear repeated everywhere I go in this world.

Two days later, now in my forties, Obama's foreign policy adviser Ben Rhodes apologized for my release taking so long (which was awkward but fulfilling), acknowledging that I was a victim of the nuclear negotiations.

But everyone knew that already. More important was the confirmation of something I had suspected for a long time. "Burma was a dress rehearsal for Cuba, and Cuba was a dress rehearsal for Iran," Rhodes told me, "although everyone understands that Iran will be much more difficult."

If only they had been able to get the nuclear deal with Iran done a couple of years sooner we might be looking at a very different world today.

"You shouldn't have filed that lawsuit. Everyone knows you were innocent. Why not just let it go?"

It was October 2016 and Yegi and I had begun journalism fellowships at Harvard. Just outside of Cambridge in Watertown, Massachusetts, we discovered a solid Iranian takeaway place, and the owner made this very simple observation after recognizing my name from my credit card.

It was the clearest and most concise argument I'd heard against the decision I made to take the government of Iran to court for their imprisoning me.

All the other points that could be made supporting or opposing my decision seem like fluff.

Some say my suing Iran raises the odds that other dual nationals will be taken hostage there. That's unlikely, because Iran is already picking up many people from a range of nationalities, at least a dozen of them since my release alone. Eighty-year-old Baquer Namazi, a former UNESCO executive who was arrested when he returned to Iran to fight for his son Siamak, who was detained when I was already in Evin. Or Xiyue Wang, a scholar at Princeton University who studied Iranian history and was arrested on espionage charges for scanning documents at the national archives that were over a hundred years old. Or most frustratingly of all, Nazanin Zaghari-Ratcliffe, a dual UK-Iranian national who was arrested while visiting her parents in Iran with her two-year-old daughter, Gabriella. It's now been nearly three years since she, her little girl, and her husband, Richard, were forced apart.

Each new farcical arrest is a reminder that taking hostages, fifty-two of them in fact, was the signature move of this regime when it first started forty years ago.

Friends and relatives in Iran worry about possible retribution against themselves, but I remind myself, *When I was in prison, where the hell were they?* Then I remember, *They were at home living their lives.*

The same thought runs through my mind every time an Iranian, or anyone of any background for that matter, who has never had a day of their lives violently stolen from them tells me that suing the Islamic Republic will somehow have a negative impact on how that country is viewed in the world. After a decade of trying to cover Iran in the most neutral tones, often giving it the benefit of the doubt, I know that nothing is better at sullying its reputation than the Iranian system itself.

Then there are those who say that by pursuing this lawsuit I am attempting, unfairly, to gain Iranian assets or force American taxpayers to pay for my hardship. They say that anyone who goes to Iran should know what they're getting into and therefore is responsible for what happens to them there.

And I can imagine my interrogators saying, "What about the time I brought him some peanuts? We weren't that bad." From the depths of my soul I can say, "I agree. Kinda." They weren't *that* bad.

Of course, they were horrible, but they could have been worse. They were human and that came through sometimes.

But for the less litigiously seasoned among us, when filing a legal complaint it's not the plaintiff's job to make the other guy look good.

A strange trait of our species is that we often say the same things. There are two concepts that get repeated so often when I get recognized that I think someone is fucking with me.

Absolutely nothing gets under my skin more. Usually it's a middle-aged guy, often slightly older than me. He'll bow his head low, sometimes put an arm around my shoulder, and say in hushed tones, "Did they beat you? Rough you up pretty bad?"

Sometimes I dodge and deflect. Other days I will explain that, no, fortunately I was spared from the actual physical violence.

And the follow-up is invariably, "Good, so they didn't mistreat you."

I hear the echoes of my captors in their dumb voices. *Isn't it usually better just to keep your mouth shut?*

"No," I tell them. "Besides my wife and I being abducted from our home at gunpoint, blindfolded, taken to prison and thrown in solitary confinement, interrogated relentlessly for several months, and denied due process; besides my being subjected to a secret closed-door trial while vicious lies were being spread about me by corrupt officials and through the local media, separated from my wife for a year and a half, forced to live in rooms where the light was always on, deprived access to information and the right to defend myself, and having the livelihood I took years to build stolen from me . . ." Then I pause. "Besides all that it was great."

It's in those moments that I'm certain I'm doing the right thing.

I've tried to let so much of what happened during those 544 days fade out of my day-to-day thoughts, but sometimes I think about those last hours in Evin when Borzou and Kazem were waiting to take me to the airport—while their comrades plotted keeping Yegi in Iran—our conversation turned to the 2016 election.

"While we still have you, explain to us about this Electoral College," Borzou said.

"Seriously?"

"Yes, it makes no sense to us."

"It doesn't make much sense to us either," I confided, then gave my best attempt at a high school civics explanation of our electoral process in Persian.

"But what if you want the popular vote to count and not this silly system? Each state can choose," Borzou said, "isn't that what is meant by states' rights?"

Like so many of their countrymen, they didn't know anything about American politics other than they were sure that they were much more democratic than their own.

It was a rare opportunity to use one of their favorite responses on them: "That's not how our American system works."

"But we want to know who you think will win. Clinton or Trump?" Borzou asked.

I didn't want to answer. *Was it some kind of trick? Did it even matter?*

"J thinks Clinton will win," Kazem announced. It wasn't the first time he had *mis*spoken for me.

The campaign was just picking up steam. Iran's state television was covering Clinton and Trump as the obvious nominees.

"No, I don't think either one of them will win," I said.

"Why not?"

"Too many people don't like her, and he's an idiot," I said, offering the most concise analysis I could provide.

"Who do you guys think will win?" I asked. I was genuinely curious.

"Our analysis is that Trump will become president," Borzou announced.

I couldn't help laughing. "You guys are even dumber than I thought," by that point it had been many months since I held back any of my contempt for these two. "How did you come to that conclusion?"

"It's very simple. Trump is the candidate that hates Muslims most," Borzou explained.

It was my first tiny preview of the future into which I was stepping.

Acknowledgments

A story like this, told from the author's point of view, may seem to the reader like a solitary endeavor. That's not the case. This book would have been impossible for me to write—for so many reasons—without the input of the following people and many others.

My employers and colleagues at the *Washington Post* have been as generous to me in contributing their insights for this book as they were loyal and fierce in their struggle to win my freedom.

Owner Jeff Bezos, publisher Fred Ryan, and executive editor Marty Baron all shared sketches of conversations and encounters they had during my ordeal. Doug Jehl, my boss when I was hired at the *Post* and my editor, has spent long hours with me since my release—talking on the record and informally—about those agonizing eighteen months and the many twists and turns in the story. He's been a supportive friend to Yegi and me as we transition to our new life in Washington, DC.

In the early weeks after my release Bob Woodward stepped forward, making himself available as an adviser to me, helping me formulate ideas at a time in my life when that seemed impossible. He invited me into his home and counseled me in every aspect of getting my experience from memory to page, talking with me, and drawing out key elements of the story. When I stop and ponder *that* I am reminded of just how fortunate I have been.

His assistant, Evelyn Duffy, transcribed those conversations, which are the basis for many of the action-oriented sections of the book. Just as crucially, Bob convinced me that—despite the many people who were telling me I had to go on television to tell my story—this was my story to tell and I shouldn't let anyone else do it for me. I am forever grateful for his wisdom, guidance, and friendship during a chaotic time in my life.

My best friend at the *Post*, Tiffany Harness, who had encouraged me to tell the sorts of stories I wanted to from the moment I joined the paper, spent long hours during days off to do audio interviews with me that helped me to identify some of the more compelling bits of my experience. She continues to be the ideal friend and colleague.

Tracy Grant, the *Post*'s Managing Editor for Staff Development and Standards, held my hand through the very long road to returning to work. It wasn't the easiest journey, but both of us have been through tougher, and I appreciate her honesty and loyalty in guiding me through the process.

I owe a debt of gratitude to Fred Hiatt, Ruth Marcus, and Jackson Diehl for giving me the opportunity to join the *Post*'s Opinions team, and to my Global Opinion colleagues, Eli Lopez, Karen Attiah, and Christian Caryl, for providing me the space to write about things I understand and am passionate about.

I want to pay special thanks to *Post* cartoonist Michael Cavna for his living sketch of my imprisonment, updated daily, a grim reminder to *Washington Post* readers of my plight, and Carol Morello, a national security correspondent, who gracefully wrote so many of the stories about an imprisoned colleague. Anyone would be lucky to call Michael and Carol their friends.

David Rohde, the great investigative reporter and now editor, who had his own hard-to-imagine experience of being held hostage, has been my postcaptivity Sherpa, expertly guiding me to something like a normal life. Since our first conversation in March 2016, speaking to him has felt like being home. I don't know how else to describe it.

One of the things I'm most grateful to David for was the introduction he made to Hostage US and its founder, Rachel Briggs. If you read this book and asked yourself, "What can I do?" I've got a good answer. Support this organization. Hostage-taking is a crime with such far-reaching implications, but often lost in the shuffle are the people—hostages and their families—that are impacted. Rachel and her deputy, Liz Frank, expertly advocated for Yegi and my recovery in every possible sense of the word, including introductions to healthcare providers, navigating insurance claims, financial planning, and so much more. It took almost three years, but together we were finally able to get to the right person at the IRS who agreed that, yes, being held hostage by a foreign government is good reason to forgive penalties on late tax returns. Seriously, there is so much that goes into rebuilding one's life after such an ordeal and Hostage US has been integral in that project.

The community of press freedom organizations, their staffs and extended communities, were among our best advocates and I want to specifically acknowledge Joel Simon, Executive Director of the Committee to Protect Journalists, and Delphine Halgand, the former

North America Director of Reporters Without Borders, for their dedication to initiatives that made a difference in my fate.

Endless appreciation goes to Bill McCarren, Executive Director of the National Press Club, who has been the source of so much support in Yegi and my quest to build a new life in Washington, DC. Before we had ever met him he had organized so many events to draw attention to our plight; it was Bill who reached out to Muhammad Ali, securing the Champ's statement and coordinating its release for maximum public impact.

"Uncle Bill," as he's known at our house, has been one of our guardian angels, ferrying us through the difficult early months of reentry. He made key suggestions and introductions that led to so many other opportunities.

First of these was connecting us with Steve Knapp, former president of George Washington University. After a very short meeting in April 2016, President Knapp offered Yegi and me a home on GW's campus. That early nineteenth-century construction on 21st Street, known as the Lenthall House, became our first address in the capital. In its basement is where most of this book was written, and where we watched many Golden State Warriors games very late into the night.

Dr. Knapp and his chief of staff, Barbara Porter, also introduced us to Frank Sesno, the director of GW's School of Media and Public Affairs. Frank offered me a fellowship that provided me the time and space to write, but also the opportunity to inch my way— at my own pace—into public life. His encouragement and guidance is a gift that keeps giving. The students at GW, and my interactions with them while we lived on that campus, are a highlight of this period in my life. Bruce Terker, who generously funds that fellowship, has become a friend to us and the future of good journalism.

Bill McCarren also encouraged me to pursue the Nieman fellowship at Harvard, which turned out to be one of the best things that has ever happened to Yegi and me.

The Nieman Foundation, and the fellowship's curator, Ann Marie Lipinski, were integral in giving us an ideal place and time to reintegrate into an intellectual environment that was simultaneously a period of healing and normalization. The Nieman staff and our cohort of fellow fellows and their families will always hold special places in our hearts for their friendship and support. And special thanks to Steve Almond, who taught a nonfiction writing class at the Nieman fellowship in which I wrote a segment of this book.

The legal team at WilmerHale—whom I have written sparingly about in this book—played an outsized role in bringing me home. The team of lawyers working on our behalf was so extensive that it still makes me blush, but Bob Kimmitt, Dave Bowker, Maury Riggan, Rob McKeehan, Jess Leinwand, and Lorraine Marshall have all become friends that will be with us forever.

As Bob has told me on several occasions, if you "ever get wrongfully imprisoned abroad, you'd want the *Washington Post* as your employer, Mary, Ali, and Yegi Rezaian as your family, and WilmerHale representing you." I concur.

Several government officials were generous with their time and insights. Those conversations have helped me be more confident about my understanding of why I had to endure what I did, and the story I told in these pages. They include John Kerry, Brett McGurk, Wendy Sherman, Rob Malley, Sahar Nowruzadeh, Ben Rhodes, Valerie Jarret, Jon Finer, Katie Ray, Jenny Farar, and others.

I have to call out Congressmen Dan Kildee and Jared Huffmann, who represented Amir Hekmati and me in the Capitol during our captivity. They met with our families, members of the Obama administration, and visiting Iranian officials and managed to make time to represent the Democrats in the annual Congressional baseball game. They used practices to strategize on our freedom campaigns; the ultimate double-play combination. Jared and Dan also came to Germany to welcome us to freedom, which my family will never forget.

To others who may prefer not to be named, please know how much I appreciate you, your service to this country, and all you did to ensure my safe return.

A handful of trusted friends—Georg Diez, David Lang, Laura Secor, Chris Schroeder, and Erick Sanchez—read drafts of this book at different stages. Your feedback and encouragement was invaluable, and I hope you see some of the changes and additions you suggested reflected in the final version. Charlie Rentschler, Charles-Antoine Joly, Mahdis Keshavarz, and Marshall Tuck all listened and gave me solid advice when I needed it most.

While I was locked up and could only call up memories, doing so brought me much-needed mirth. When it was possible, I shared those stories with Mirsani and Yadoallah. I miss the two of them tremendously and owe my survival and sanity to their comradery.

This book would not have been possible if not for Anthony Bourdain. First and foremost, his endless advocacy while I was in prison was essential in raising the international awareness of our ordeal. That commitment remained consistent and continued well after our release and until the end of his life. People know of our connection from Yegi and my appearance on *Parts Unknown*, but it went so much deeper than that and our lives are immeasurably richer for having known him.

Tony also introduced me to my agent, Kim Witherspoon. In all frankness, I would be lost without her and am fortunate to have her representing me. I hope I am becoming the author she thought I could be. That she stood with me while I was still wobbly from my time in captivity is the source of endless gratitude.

William Callahan, Emme Schlee, Jessica Mileo, and Maria Whelan—members of Kim's team at Inkwell Management—were always available when I needed them.

Once Kim, William, and I had prepared what we thought was a strong book proposal, Kim took it to publishers. Several were interested, including Tony, who made a strong case for me to consider Anthony Bourdain Books at Ecco as the home for this story.

From our first conversation, Ecco's publisher, Dan Halpern, has been committed to Tony's vision of me telling the story I wanted to tell, and I believe that's what I've done. In that process my editor, Zack Wagman, has been a faithful partner who has made suggestions and edits that consistently make me look like a better writer than I actually am. Book writing is a collaborative process and Zack has been the ideal comrade in bringing this one to life.

Finally, without my family—specifically my wife, brother, and mother—not only would it have been impossible for me to write this book, I'd also, in all probability, still be in prison. Your commitment, generosity, and loyalty to me means everything. Hopefully those are attributes that I can live up to in a fraction of the way the three of you have shown up for me.